SUCCESSFUL SOLAR ENERGY SOLUTIONS

3

SUCCESSFUL SOLAR ENERGY SOLUTIONS

Spruille Braden III · Kathleen Steiner

with ALVIN O'KONSKI

VNR **VAN NOSTRAND REINHOLD COMPANY**
NEW YORK   CINCINNATI   ATLANTA   DALLAS   SAN FRANCISCO
LONDON   TORONTO   MELBOURNE

Van Nostrand Reinhold Company Regional Offices:
New York   Cincinnati   Atlanta   Dallas   San Francisco

Van Nostrand Reinhold Company International Offices:
London   Toronto   Melbourne

Library of Congress Catalog Card Number: 78-26123
ISBN: 0-442-20738-7

Manufactured in the United States of America

Published by Van Nostrand Reinhold Company
135 West 50th Street, New York, N.Y. 10020

Published simultaneously in Canada by Van Nostrand Reinhold Ltd.

15 14 13 12 11 10 9 8 7 6 5 4 3 2 1

Library of Congress Cataloging in Publication Data

Braden, Spruille, 1953-
   Successful solar energy solutions.

   Bibliography: p.
   Includes index.
   1. Solar heating. 2. Solar air conditioning.
I. Steiner, Kathleen, joint author. II. O'Konski,
Alvin R., joint author. III. Title.
TH7413.B73 697 78-26123
ISBN 0-442-20738-7

Dedication

Dr. Forrest Wilson, Ph.D., a carpenter, sculpter, seaman, ship's carpenter, construction superintendant, architect, designer, interior designer, chairman and associate dean of architecture schools, philosopher, writer, husband and father, who revives in those surrounding him the moral obligations inherent to their positions in life.

He is our mentor and friend.

S.B. III
K.L.S.

Acknowledgments

This investigation was made possible in part by a grant from Van Nostrand Reinhold Company. Special thanks are due to Mr. Eugene Falken and Mr. Jean Koefoed for their receptiveness.

We owe a particular debt of gratitude to Mr. Alvin O'Konski, who developed and printed the photographs. We thank Ms. Katherine Davis for typing the manuscript, and Mr. Chris Romney for traveling with us and photographing some of the buildings.

Thanks are also extended to Mr. and Mrs. Spruille Braden Jr., Mrs. Carolyne Steiner, Ms. Barbara Royal, Ms. Izabel Manzinelli, Mr. and Mrs. Richard Steiner, Mr. and Mrs. John Holubuwicz, Mr. and Mrs. Herbert Royal, Mr. and Mrs. John Czarnecki and Mrs. Frieda Steiner. Their more than generous hospitality greatly facilitated our trip.

Although it is impossible to give credit to everyone involved, we do wish to thank the owners, architects, engineers and designers of the buildings presented here.

Introduction

Buildings have traditionally reflected the existing social trends, thinking patterns, and technical achievements of the societies in which they are created. The buildings in *Successful Solar Energy Solutions* reflect all of these parameters for an evergrowing part of our society; they reflect a drive toward self-sufficiency, an awareness of environmental issues, and numerous levels of technical achievement.

The structures were not selected to illustrate a preconceived point of view, but rather because they represent what we feel is the most innovative, positive, and well-defined direction followed by the building community in the last 20 years. The increased awareness of the financial and political limitations imposed by the high levels of energy consumption has fostered a new thinking pattern. A few of the many successful buildings produced by this new pattern are presented here. What makes these buildings successful is the way in which the spatial and mechanical concepts are combined to create exciting, comfortable, and affordable environments.

The book presents buildings that employ a wide range of solar energy systems—namely, passive and active space heating/cooling and domestic water heating. Applications of these systems are shown for both new and retrofitted buildings. The buildings are grouped into three types: institutional, residential, and commercial. In turn, samples of each type are presented from the four basic climatic regions: cool, temperate, hot humid, and hot arid. Within this structure, each building is treated as a separate case study so that the reader can compare case studies based on similarities of function and climate.

The case studies present three distinct areas of information. The first area is a listing of climatic and geographic information, as well as a listing of the individuals involved in the design and construction of the building. The second, spatial solution, illustrates how a building's spatial characteristics contribute to the collection and conservation of solar energy. The third area, mechanical solution, illustrates the operating modes and construction details of the solar energy systems. This area gives the percentages of the heating and cooling load that are provided by the solar systems. These percentages are calculated averages and will vary from one year to the next, subject to weather intensity.

The information rendered in the drawings and photographs is not covered in the text portion. Therefore, it is imperative for the reader to study the graphic presentations in each case study. Economic data are not presented because fluctuating oil and gas prices limit the period of time for which this information is valid. Finally, Appendix A contains brief descriptions of the predominant climatic regions in the United States. Appendix B summarizes solar energy systems and describes their operating modes. Appendix C lists organizations involved in furthering the implementation of solar energy systems. Appendix D lists over 500 manufacturers of solar energy products, both alphabetically and geographically.

Contents

SUCCESSFUL SOLAR ENERGY SOLUTIONS

PART I
INSTITUTIONAL BUILDINGS

Section 1. Cool Region

The buildings presented in this section are the following:

Maine Audubon Headquarters. Falmouth, Maine.

Sheet Metal Union School. Detroit, Michigan.

Federal Office Building. Saginaw, Michigan.

North View Junior High School. Brooklyn Park, Minnesota.

Mount Rushmore National Memorial Visitor Center. Keystone, South Dakota.

The Bighorn Canyon National Recreation Area Visitor Center. Lovell, Wyoming.

MAINE AUDUBON HEADQUARTERS

LOCATION: Falmouth, Maine.
LATITUDE: 44 degrees N.
REGION: Cool.
ARCHITECT: G.B. Terrien, Falmouth, Maine.
SOLAR CONSULTANT: R. C. Hill, Maine.

SPACIAL SOLUTION

This building serves as the central office facility and educational center for the Maine Audubon Society. It is open to the public as a demonstration of an alternative solution in energy technology.

The building is located on a flat and open site. Its plan is oblong in shape, with the longer sides facing directly north and south. Its volume is moulded into the traditional salt box form, which is responsive to the environment and to energy conservation. The low, north-sloping roof reduces the wind resistance on the cold side of the building, while yielding a maximum surface area facing south. The building uses a wood frame construction erected on a poured concrete slab. The total floor area is 5,500 square feet, which is divided between two levels.

Due to extremely efficient use of insulation (5-1/2 inches in the walls and 6 to 9 inches in the roof, with 2 inches of polyurethane foam around the perimeter slab), the building's heat losses are only 40 percent of the total heat load; 60 percent of the load accounts for the need to heat fresh air. There is one air change per hour. To reduce heat losses, the entrance doors use an airlock design. The windows on the east, west and north facades have a combined area of 370 square feet and are triple glazed. There are two 4-foot wide double glazed strips on the south elevation, allowing solar heat and light to penetrate into the interior spaces.

MECHANICAL SOLUTION

The Maine Audubon Society Headquarters building uses solar energy for space heating and for domestic hot water. The heating system may be considered unique in three aspects. First, it implements air-cooled, flat-plate collectors, which have been used in the New England area but have never before been installed in this climate or latitude. Second, the collectors were designed to be built and assembled on the site, from "off the shelf" components. Third, the use of a wood furnace eliminates the need for conventional energy sources for back-up and auxiliary systems.

The collector is a flat-plate, air-cooled type, tilted to 50 degrees from the horizontal. The collector area is 2,500 square feet and the ratio of floor to collector area is approximately 2 to 1. One of the main considerations for the design of this system was the reduction of the heat lost back to the outside environment. In order to achieve this, the collectors funnel the heated air away from the transparent cover and towards the back plate. The air is circulated through black-coated aluminum window screening, which is stapled to the plywood plate. For every foot of plywood, there are 3 feet of screen, made possible by moulding the screen into a continuous "S" shape. The heated air rises to the top of the collector and is returned to the storage bin with the aid of three 1-horsepower blowers, which force the air downward directly behind the plywood sheet. When desired, the heated collector air can be directly circulated to the rooms. Professor Richard K. Hill of clthe University of Maine is responsible for this unique design.

The heat storage is composed of crushed granite with an average diameter of 1 inch. The stones are contained in a galvanized steel grate. Heated air from the collectors is circulated along the vertical axis of the stone bed, rather than through the much longer horizontal axis. This reduces friction and the energy needed to circulate the air.

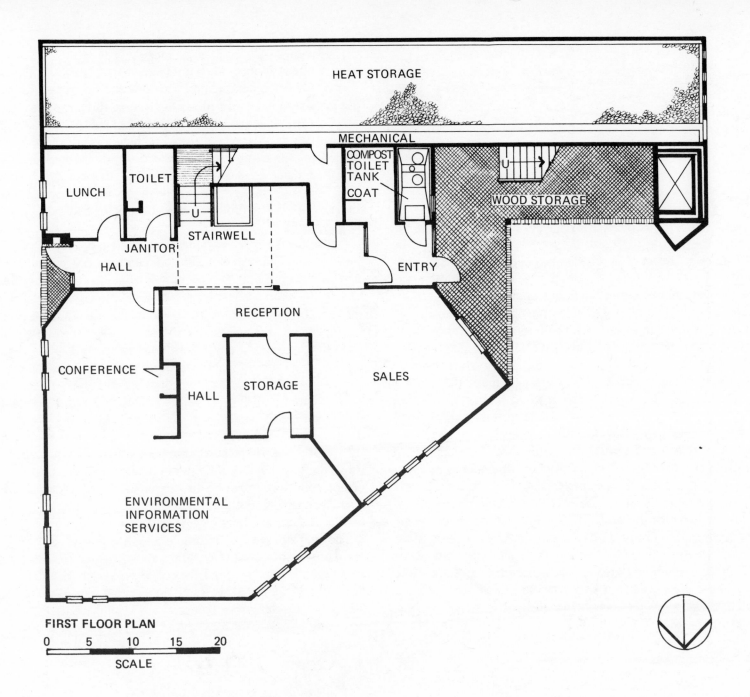

HEAT STORAGE

MECHANICAL

LUNCH

TOILET

COMPOST
TOILET
TANK

COAT

U

WOOD STORAGE

JANITOR

STAIRWELL

HALL

U

ENTRY

RECEPTION

CONFERENCE

HALL

STORAGE

SALES

ENVIRONMENTAL
INFORMATION
SERVICES

FIRST FLOOR PLAN

0 5 10 15 20

SCALE

6

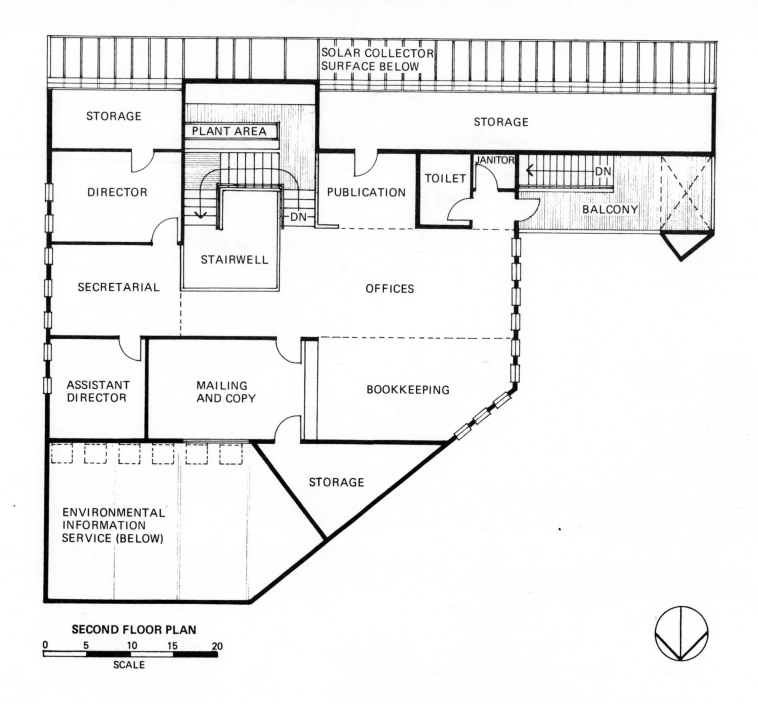

STORAGE

PLANT AREA

SOLAR COLLECTOR
SURFACE BELOW

STORAGE

DIRECTOR

JANITOR

TOILET

PUBLICATION

DN

STAIRWELL

BALCONY

SECRETARIAL

OFFICES

ASSISTANT
DIRECTOR

MAILING
AND COPY

BOOKKEEPING

STORAGE

ENVIRONMENTAL
INFORMATION
SERVICE (BELOW)

SECOND FLOOR PLAN

0   5   10   15   20

SCALE

7

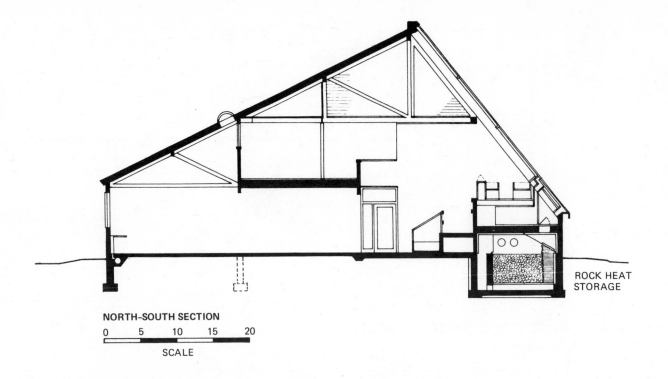

NORTH–SOUTH SECTION

ROCK HEAT STORAGE

0   5   10   15   20

SCALE

The heat distribution system is a conventional forced air design. Rooms are individually zoned to allow heating flexibility and energy conservation in unoccupied rooms.

The solar system was designed to provide 60 to 70 percent of the total heating needs and is backed up with a wood furnace that is integrated into the heat storage unit. The air pollution problems generally associated with this type of furnace have been eliminated by the use of a complete combustion cycle which consumes the wood fibers. The need to constantly fuel and regulate the stove's output is also eliminated. This was achieved by coupling the furnace with the heat storage unit used by the collectors.

It is estimated that 3 cords of wood per year will be needed to supplement the solar system. Since Maine is 90 percent forest land, which is higher than any other state in the country, wood is a feasible source of heating fuel. It was estimated that 6,800,000 cords of surplus harvestable wood were available in 1976 for home fuel needs in Maine. This is equivalent to 14,068,942 barrels of oil.

In the summer, cool air is circulated through the storage bin during the night. During hot summer days, the room air is circulated through the same bin to cool the air which is then redistributed to the rooms.

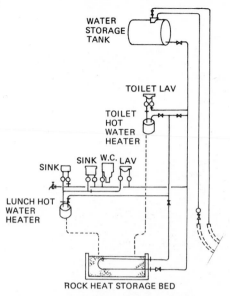

WATER STORAGE TANK

TOILET LAV

TOILET HOT WATER HEATER

SINK

SINK W.C. LAV

LUNCH HOT WATER HEATER

ROCK HEAT STORAGE BED

—O— STOP VALVE AT FIXTURE
—|— CHANGE IN PIPE TYPE OR DIAMETER
—▷◁— STOP AND WASTE VALVE
—O▷◁— GATE VALVE

PLUMBING RISER DIAGRAM SCHEMATIC

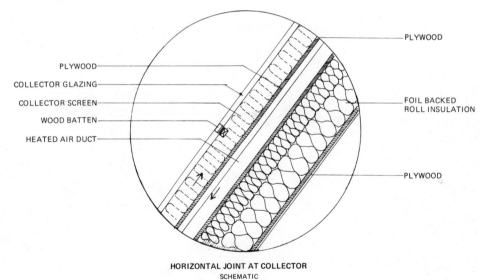

PLYWOOD

PLYWOOD

COLLECTOR GLAZING

COLLECTOR SCREEN

WOOD BATTEN

HEATED AIR DUCT

FOIL BACKED ROLL INSULATION

PLYWOOD

HORIZONTAL JOINT AT COLLECTOR
SCHEMATIC

Southwest elevation

Interior skylight

Northwest elevation

10

SHEET METAL UNION SCHOOL

LOCATION: Detroit, Michigan.
LATITUDE: 42 degrees N.
REGION: Cool.
ARCHITECTS, ENGINEERS AND SOLAR
   CONSULTANTS: OEM Associates of St. Clair Shores,
Michigan.

SPACIAL SOLUTION

This building is a school for apprentice sheet metal contractors. It is located in a commercial district of a residential area a few miles from downtown Detroit.

The school's exterior architecture is well-integrated with the surrounding environment. Its construction is masonry. The one-story, 10,360-square-foot building has the main entrance from the south and is recessed to achieve some protection from winter winds. There are very few windows, most of which are located on the west elevation.

MECHANICAL SOLUTION

The school is in itself a laboratory for innovative approaches and developments in the field of solar heating. The heating design can best be described as a solar assisted heat pump system.

The collector is a flat-plate, air-cooled type produced by Solaron of Denver, Colorado. The building has 3,104 square feet of collectors facing directly south. The ratio of floor to collector area is approximately 3 to 1.

The storage medium is 28 tons of rock stored in a 1,560-cubic-foot bin located above ground. The reason for this placement is to allow future improvements to take place without major disruptions to the system. The ratio of collector square feet to storage cubic feet is approximately 2 to 1.

The school's heating system enjoys an innovative design in which the back-up element is so well-integrated into the over-all system, it can no longer be considered an auxiliary element and must be viewed as a component of the primary system. The heated air from the collector is circulated through the storage bin, and from there it is directed by the air handling unit to the nine heat pumps. If additional energy is needed to raise the air

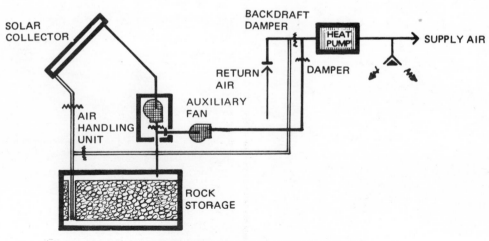

SOLAR ASSIST HEAT PUMP SCHEMATIC

temperature, the heat pumps inject the heat. There is no conventional fuel back-up system.

The solar system provides 50 percent of the heating requirements. The remaining half is made up by the heat pumps, which also provide space cooling.

Southwest elevation

FEDERAL OFFICE BUILDING

LOCATION: Saginaw, Michigan.
LATITUDE: 43 degrees, 39′ N.
REGION: Cool.
FINAL DESIGN: U.S. General Services Administration, Chicago, Illinois.
INSTRUMENTATION: University of Michigan.

SPACIAL SOLUTION

This building was awarded a design citation in the 1974 annual Progressive Architecture Awards Program and was the winner of the governmental category in the 1974 Owens-Corning Fiberglass Energy Conservation Awards Program.

The federal office building provides office space for thirteen federal agencies. In addition to housing a post office, a loading dock and parking space for 100 official cars are provided.

The building is one story with 59,000 square feet of floor area. The walls and roof are bermed for better insulation and to create a recreational park. The main entrances are from the east and west and are recessed to provide protection against winter winds. The building makes use of large double glassed windows inserted into eaves with overhangs, which keep the sun's rays from directly entering the interior spaces. Numerous energy and water saving features are implemented throughout.

MECHANICAL SOLUTION

The project involved the fabrication and installation of a large solar collector array which will serve as a sample for future buildings with similar functional requirements.

There are 8,000 square feet of tubular, vacuum-jacketed, water-cooled collectors tilted to 45 degrees from the horizontal and mounted on pre-cast, "T"-shaped concrete beams. The ratio of floor to collector area is approximately 7 to 1.

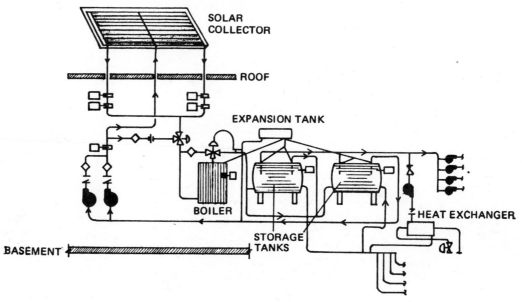

SOLAR COLLECTOR

ROOF

EXPANSION TANK

BOILER

STORAGE TANKS

HEAT EXCHANGER

BASEMENT

SOLAR SYSTEM SCHEMATIC

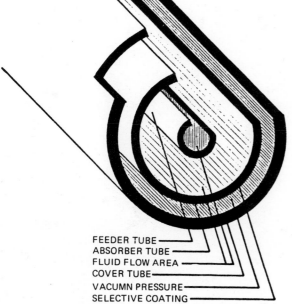

FEEDER TUBE
ABSORBER TUBE
FLUID FLOW AREA
COVER TUBE
VACUMN PRESSURE
SELECTIVE COATING

SCHEMATIC DETAIL OF VACUMN TUBE COLLECTOR

The heat absorbing medium is water, with no antifreeze or other chemicals added. The water is circulated through the system by a 5-horsepower centrifugal pump, and it flows directly from the collector to the storage unit without the use of a heat exchanger.

The storage is composed of two 15,000-gallon containers. The heated water is circulated to the rooms by perimeter baseboard radiators, with a boost from fan-coil units. The back-up system is a gas fired boiler.

An absorption chiller provides the building with space cooling. The solar system provides 60 percent of the space heating, 100 percent of the domestic hot water and 60 percent of the space cooling requirements.

13

Southwest elevation

West entrance and collector support

North elevation

14

Windows, north elevation

NORTH VIEW JUNIOR HIGH SCHOOL

LOCATION: Brooklyn Park, Minnesota.
LATITUDE: 45 degrees N.
REGION: Cool.
MAIN SOLAR ENGINEERING AND
    CONSTRUCTION: Honeywell, Inc., Illinois.

MECHANICAL SOLUTION

The school is located in a northern suburb of Minneapolis. The solar system does not form an integral part of its architecture.

Energy conservation measures in this building are limited to the implementation of the solar heating system.

The 5,000 square feet of flat-plate, liquid-cooled collectors are located to the southeast of the building and tilted to 55 degrees from the horizontal. They use a combination of water and ethylene glycol as the heat removing medium.

The storage element contains a heat-exchanger and has a capacity of 3,000 gallons. The system provides 54 percent of the space heating requirements and 21 percent of the domestic hot water, which includes the heated water for the swimming pool.

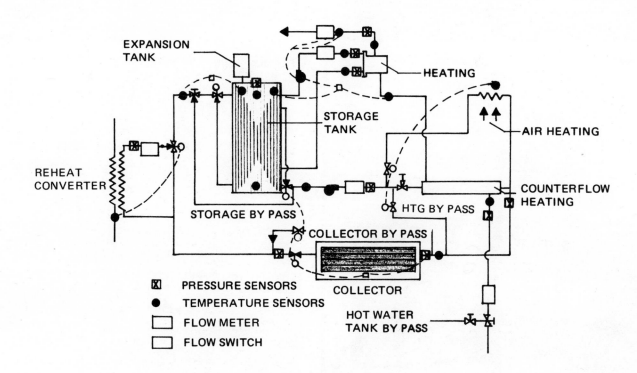

SOLAR SYSTEM SCHEMATIC

Site mounted collectors

Northwest elevation

MOUNT RUSHMORE NATIONAL MEMORIAL VISITOR CENTER

LOCATION: Keystone, South Dakota.
LATITUDE: 43 degrees, 50′ N.
REGION: Cool.
ARCHITECT AND SOLAR CONSULTANT: The Spitzanel
    Partnership, Inc. Work was done under a contract with the
    South Dakota School of Mines and Technology of Rapid
    City, South Dakota.

SPACIAL SOLUTION

The original concept of using the Center as a demonstration site for solar energy originated with the park officials and was based on three major items. First, for solar energy, the area enjoys a high number of sun days per year; second, the building's siting gave the proper orientation for retrofitting it to utilize solar energy; and third, the number of people visiting the center has reached 2,250,000 per year. Therefore, this presented an excellent opportunity for vast exposure to the public of an operating solar system.

The building has 9,250 square feet, the totality of which is heated by the solar system, although cooling is limited to the observation room, which has 2,000 square feet in area.

MECHANICAL SOLUTION

This is a retrofit project, with the solar system making use of the existing heating and cooling mechanism for back-up purposes.

The collector is a flat-plate, liquid-cooled type that uses a combination of water and antifreeze as the heat removing medium. There are 2,000 square feet of collectors mounted on the roof facing southwest. The ratio of floor to collector area for the heating mode is approximately 4 to 1; for the cooling mode, it is 1 to 1.

The heated liquid is circulated from the collector to the heat exchanger, where the thermal energy is transferred to the storage element. The insulated storage tank has a 3,000-gallon capacity.

Southwest elevation

For space heating, water from the storage tank or the heat exchanger is pumped through coils in the existing forced air system. The storage is bypassed when the building can use the heat directly from the collectors. For cooling, the heated liquid is used to power the refrigerant element, which also has coils in the forced air system. The system provides 53 percent of the heating and 41 percent of the cooling requirements for the center.

The previously installed heating and cooling mechanism is used when the collected thermal energy is insufficient to produce the desired results.

THE BIGHORN CANYON NATIONAL RECREATION AREA VISITOR CENTER

LOCATION: Lovell, Wyoming.
LATITUDE: 45 degrees N.
REGION: Cool.
ARCHITECT: Wirth Associates, Lovell, Wyoming.
ENGINEER: Jack S. Gordon, Cody, Wyoming.
SOLAR CONSULTANT: G. O. G. Löf, Chicago, Illinois.

SPACIAL SOLUTION

Although many would consider this building to be located in the middle of nowhere, the project is living proof that the implementation of solar energy, in locations where hook-ups to conventional power sources would be difficult and expensive, can be done with ease and at the right price. The site may be described as undeveloped and arid. The building is "L"-shaped in plan, with the longer leg facing directly south. Its section is formed like the traditional salt box house. The construction is masonry, with steel studs for interior partitions. The Center is a one-story building with 9,500 square feet of floor area. The main entrance is deeply recessed and flanked by two walls, to provide protection against the summer sun and winter winds while creating an aesthetically pleasing transition space. There are approximately 120 square feet of windows in the north facade, and they are slightly recessed. The east and west elevations have smaller glass areas, angled to reduce direct heat gains. There are no windows in the south facade.

MECHANICAL SOLUTION

The park service uses the solar system as the primary heating and cooling source.

The 2,500 square feet of flat-plate collectors are air-cooled and form an integral part of the roof. The collectors are tilted to 48 degrees from the horizontal. The panels are manifolded side by side to minimize duct work and roof penetrations. Air is circulated through the system by a 5-horsepower blower. The energy harnessing capacity of the system is increased by a small reflective pond located south of the collectors.

The ratio of floor to collector area is approximately 4 to 1. The storage element is 75 tons of stones, and the ratio of collector area to storage volume is approximately 1.6 to 1.

During the summer months, the stones are cooled by circulating cool night air through the bin. During the day, this cool air is blown to the rooms, and a spray of water is used to increase the cooling effect.

The system provides 70 percent of the heating requirements and 60 percent of the cooling requirements. The back-up system is gas fired.

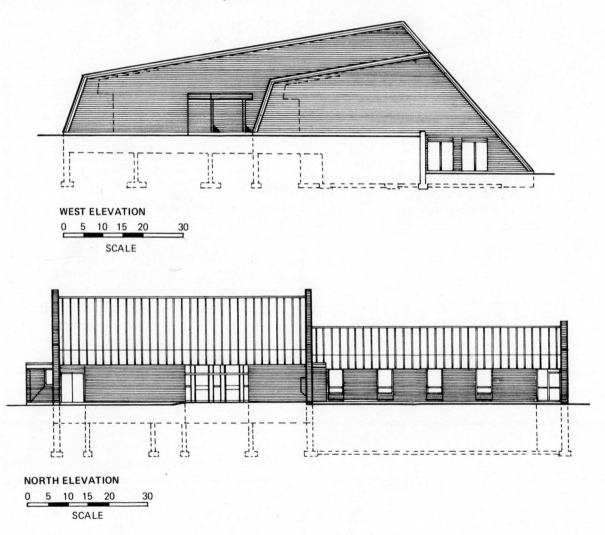

WEST ELEVATION

0 5 10 15 20   30

SCALE

NORTH ELEVATION

0 5 10 15 20   30

SCALE

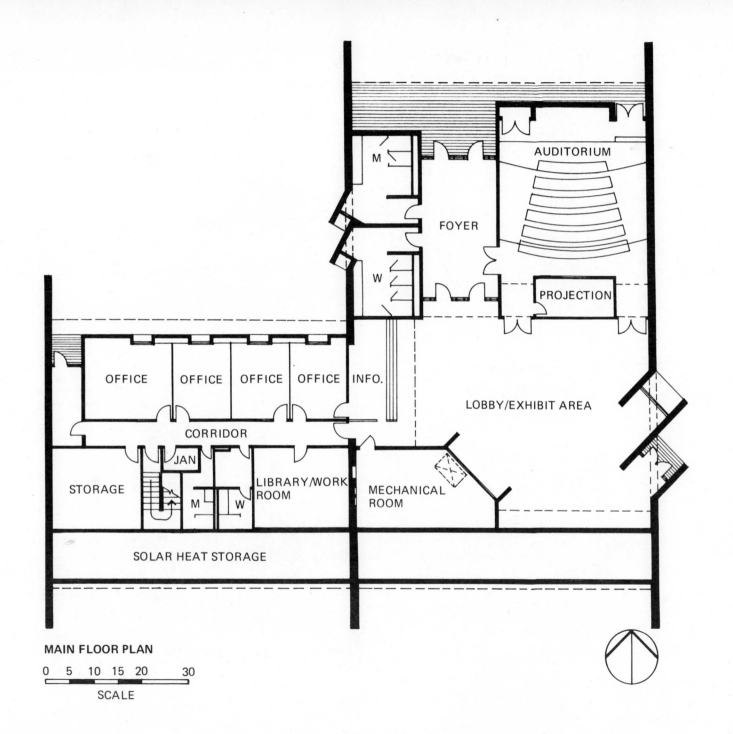

AUDITORIUM

FOYER

M

W

PROJECTION

OFFICE OFFICE OFFICE OFFICE INFO.

LOBBY/EXHIBIT AREA

CORRIDOR

JAN

STORAGE

M W

LIBRARY/WORK ROOM

MECHANICAL ROOM

SOLAR HEAT STORAGE

MAIN FLOOR PLAN

0 5 10 15 20   30

SCALE

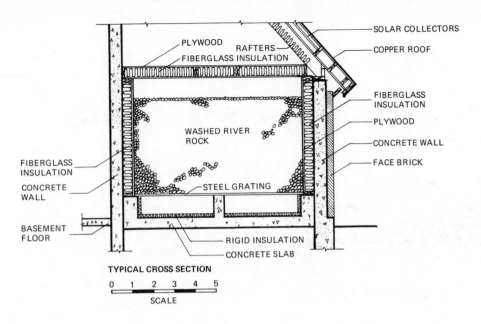

PLYWOOD
RAFTERS
FIBERGLASS INSULATION

SOLAR COLLECTORS

COPPER ROOF

FIBERGLASS
INSULATION

PLYWOOD

CONCRETE WALL

FACE BRICK

WASHED RIVER ROCK

FIBERGLASS
INSULATION

CONCRETE
WALL

STEEL GRATING

BASEMENT
FLOOR

RIGID INSULATION

CONCRETE SLAB

TYPICAL CROSS SECTION

0 1 2 3 4 5

SCALE

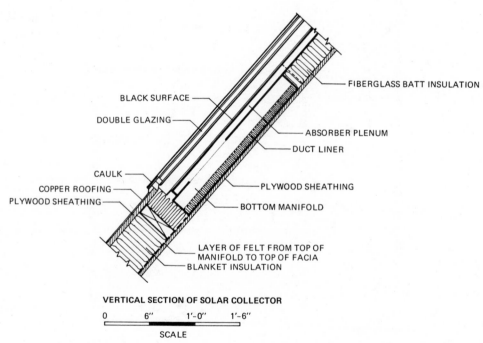

FIBERGLASS BATT INSULATION

BLACK SURFACE

DOUBLE GLAZING

ABSORBER PLENUM

DUCT LINER

CAULK

PLYWOOD SHEATHING

COPPER ROOFING

PLYWOOD SHEATHING

BOTTOM MANIFOLD

LAYER OF FELT FROM TOP OF
MANIFOLD TO TOP OF FACIA

BLANKET INSULATION

VERTICAL SECTION OF SOLAR COLLECTOR

0   6"   1'-0"   1'-6"

SCALE

20

Section 2. Temperate Region

The buildings presented in this section are the following:

The Pitkin County Airport Terminal. Aspen, Colorado.

Bus Maintenance Facility. Denver, Colorado.

Pagosa Fire Station. Pagosa Springs, Colorado.

Parker Area Junior High School. Parker, Colorado.

Christian Reform Church Center of Hope. Westmingster, Colorado.

Community College of Denver, North Campus. Westmingster, Colorado.

Armed Forces Reserve Center Armory. Norwich, Connecticut.

Upton Multi-Purpose Center. Baltimore, Maryland.

Padonia Elementary School. Cockeysville, Maryland.

Massachusetts Audubon Society Nature Center. Lincoln, Massachusetts.

Cambridge School. Weston, Massachusetts.

Norris Cotton Federal Building. Manchester, New Hampshire.

Troy Miami County Library. Troy, Ohio.

Ridley Park Post Office. Ridley Park, Pennsylvania.

Terraset Elementary School. Reston, Virginia.

Faquier County Public High School. Warrenton, Virginia.

THE PITKIN COUNTY AIRPORT TERMINAL

LOCATION: Aspen, Colorado.
LATITUDE: 39 degrees N.
REGION: Temperate.
ARCHITECT: Copland, Hagman & Yaw, Ltd., Aspen, Colorado.
SOLAR ENGINEER: Zomeworks, Inc., Albuquerque, New
   Mexico.

SPACIAL AND MECHANICAL SOLUTION

This airport terminal is one of the largest applications of passive solar heating in the United States. Due to the efficient integration of the passive solar heating components with the building's space defining elements, it is impossible to differentiate between the spacial and the mechanical solutions; therefore, the energy gathering and conserving modes are presented in unison.

The terminal is located on an open site, five minutes from the resort area of Aspen. The 16,800 square feet of floor area is used for airline offices, ticket counters, waiting rooms, cafeterias, baggage areas and various other functions associated with an airport; all of these are located on one floor. The building is composed of three connecting rectangular volumes, each facing 15 degrees east of south. Two of the volumes measure 70 feet by 70 feet in plan. The third one, which is located in the middle, measures 70 feet by 100 feet.

The wall is constructed of 8-inch blocks with concrete filled voids. The floor slabs and walls were designed to increase the building's thermal mass. The over-all R value for the walls and the ceiling is 20. The east walls are bermed to reduce heat losses. The slab and the walls are the collector and storage elements. The terminal makes use of beadwalls and skylids to control the penetration of sunlight into the interior space. The beadwalls are made of two polyester, reinforced, fiberglass sheets, spaced 2.7 inches apart. When the exterior temperature is lower than the building's interior temperature, the space is filled with styrofoam beads. This creates an insulating wall with a resis-

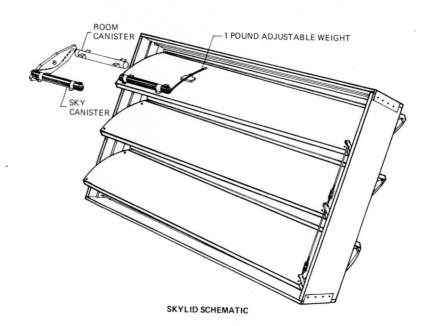

ROOM CANISTER — 1 POUND ADJUSTABLE WEIGHT — SKY CANISTER

SKYLID SCHEMATIC

tance of $R=9$. When the exterior temperature is sufficiently high to allow the slab and walls to absorb thermal energy, the beads are vacuumed out of the wall and stored. The circulation of the beads in and out of the walls takes about three minutes and is powered by a small blower.

The skylids are moving louvers which rotate about their axis and allow solar heat to enter the building when the predetermined temperature conditions are met. Each skylid operates independently and is mounted behind two layers of filon, which act as cover plates. They are tilted to 53 degrees from the horizontal. The skylids have two connected, freon filled containers, placed on opposite sides of the lid. When the sun heats the exposed container, it causes the freon to circulate to the shaded container. This shifts the weight balance and causes the skylid to rotate into an open position. When the sun stops shining, or when the exterior temperature drops, the freon moves back to the front containers, thereby closing the skylid.

This system provides 45 percent of the required heat, with the remaining portion provided by a gas fired boiler.

23

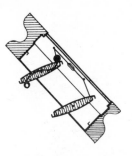

SKY CANISTER WARMER THAN
ROOM, CANISTER-LOUVERS OPEN

ROOM CANISTER WARMER-
LOUVERS CLOSE

LOUVERS TIED IN POSITION-
CANISTER SYSTEM OVERRIDDEN

SCHEMATIC OF SKYLIDS

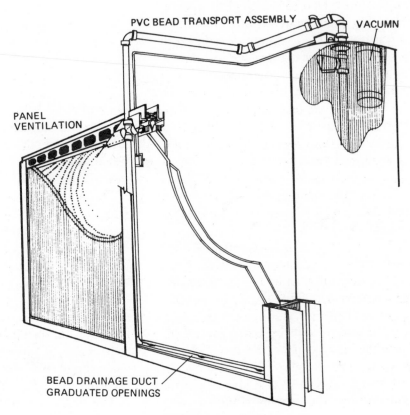

PVC BEAD TRANSPORT ASSEMBLY   VACUMN

PANEL
VENTILATION

BEAD DRAINAGE DUCT
GRADUATED OPENINGS

BEADWALL SCHEMATIC

Southeast elevation

Main entrance

Northeast elevation

Bead wall

25

BUS MAINTENANCE FACILITY

LOCATION: Denver, Colorado.
LATITUDE: 40 degrees N.
REGION: Temperate.
ARCHITECT: Charles S. Sink Associates, Denver, Colorado.
ENGINEER: Swanson-Rink Associates, Denver, Colorado.
SOLAR CONSULTANT: J. F. Kreider, Denver, Colorado.

SPACIAL SOLUTION

This building is located in a Denver suburb. Its function is to garage and provide shelter for the maintenance of 252 buses. All services are maintained on one floor, with a total area of 256,000 square feet, of which 170,000 square feet are heated. The construction is masonry and the building is oriented 45 degrees from the south. The over-all glass area is negligable. The southeast and northeast walls are bermed to within a few feet from the roof line.

MECHANICAL SOLUTION

The solar system has to provide a temperature of 40 degrees F in the bus storage area, as well as hot water to wash 250 buses every day. There are 1,394 liquid-cooled, flat-plate panels, amounting to 40,000 square feet. The panels are mounted on a steel frame located on the flat roof and tilted to 55 degrees from the horizontal. Each panel has a double cover plate and 3/4 inches of mineral wool insulation. The heat removal medium is a half-and-half combination of water and ethylene glycol, circulated through the collector and heat exchanger and powered by a 100-horsepower centrifugal pump. The 80,000-gallon capacity storage tank was built of poured concrete, insulated and placed below grade.

West corner elevation

Southeast elevation

The solar system provides 57 percent of the heating requirements. The remaining portion is provided by a gas fired boiler, with an oil furnace as additional back-up.

Cooling is provided, but the solar elements were not designed to integrate with the cooling system.

Southwest elevation

South elevation

Roof mounted collectors

Washing room

27

PAGOSA FIRE STATION

LOCATION: Pagosa Springs, Colorado.
LATITUDE: 37 degrees N.
REGION: Temperate.
ARCHITECT AND SOLAR SYSTEM DESIGNER: Hawkweed
    Group, Ltd., Denver, Colorado.

SPACIAL SOLUTION

This building's function is to house the fire engines and their related equipment, while providing a classroom in the loft area. Its 2,700 square feet are shaped into a rectangular floor plan and into a section which follows the slat box form. All the windows are double glazed and comprise less than 10 percent of the exterior surface area. The construction is wood frame, with the walls insulated to $R=19$ and the ceiling to $R=20$.

MECHANICAL SOLUTION

The solar system was designed to provide the majority of the space heating requirements.

The 700 square feet of air-cooled collectors are mounted as an integral structural part of the front roof. The black-coated corrogated aluminum plate is directly attached to the roof joists. Air is circulated in a serpentine manner from one end of the collector to the opposite end. A 1/2-horsepower blower is used to force the air through the collector and into the storage bin located at the west end.

The storage is composed of 30 tons of stones contained in a 576-cubic-foot wood frame bin, located outdoors, above ground, and surrounded with 8 inches of styrofoam insulation. The ratio of collector area to storage volume in feet is approximately 1.4 to 1. The two rooms receive heat from the forced air system, which is an integral component of the propane fired, back-up furnace. The solar system provides 95 percent of the station's heating needs.

Main entrance southwest

Northeast elevation

Southwest elevation

PARKER AREA JUNIOR HIGH SCHOOL

LOCATION: Parker, Colorado.
LATITUDE: 40 degrees N.
REGION: Temperate.
ARCHITECT: More, Combs, Burch, Denver, Colorado.
CONSULTING ENGINEER: Beckett-Harmon, Denver, Colorado.

SPACIAL SOLUTION

The architects were charged with the design and development of a 140,000-square-foot school that would be constructed in two stages: the first demanding 80,000 square feet for 600 students; the second, 60,000 square feet for an additional 600 students.

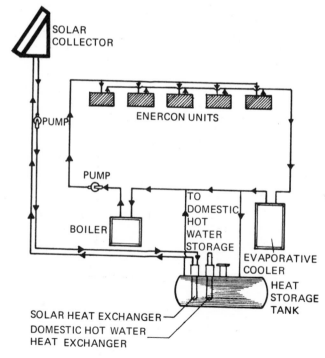

SOLAR SYSTEM SCHEMATIC

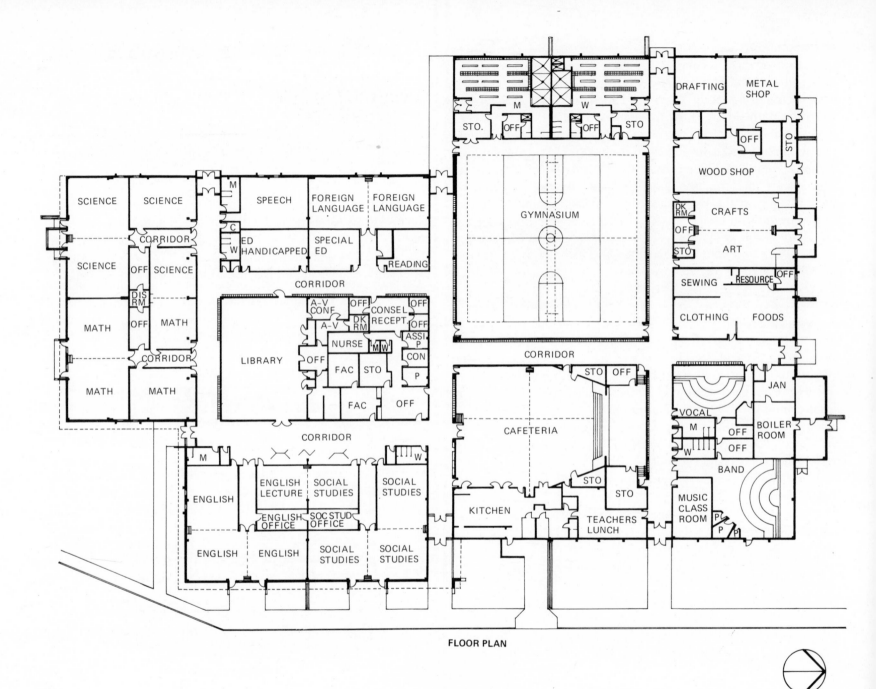

FLOOR PLAN

30

The school is located on a 14-acre site with no surrounding trees to provide wind protection. Wood fins and screens were designed into the building to reduce heat losses. The walls are constructed of 4-inch brick with a 1-inch air space, and gypsum board on both sides of fully insulated steel studs. The roof construction consists of a light weight steel deck, batt insulation, gypsum board and a suspended ceiling. Double glazed windows compose less than 12 percent of the total exterior surface area. The spacial and functional designs were developed based on their impact on the building's energy consumption and retention. The interior space was increased by 25 percent over the standard double loaded school plan corridor, which increases interior heat gain. This heat is captured by the recovery system and transferred to the storage tank.

MECHANICAL SOLUTION

The architects were encouraged to use a solar assisted system because the price of local fuel oil was 2-1/2 times more than that of gas, and the price of electricity 4-1/2 times more than that of gas. However, gas was limited by the Public Service Company to a maximum of 7,500 cubic feet per hour. Since the gas demand made by 1,200 students was estimated to be 11,200 cubic feet per hour, The Parker Area Junior High School became Colorado's first solar heated school.

There are 3,000 square feet of liquid-cooled, flat-plate collectors mounted on wood frames and tilted to 53 degrees from the horizontal. The heat removing agent is 100 percent Dow Therm "J" and it flows from the collector to the heat exchanger in the storage tank.

The storage is located below grade and has a capacity of 12,000 gallons. It provides energy for the hydronic heat pumps and preheats the domestic hot water. Cooling is provided by reversing the cycle in the pumps. A gas fired boiler with a 5,600-cubic-feet-per-hour consumption is used for back-up energy.

CHRISTIAN REFORM CHURCH CENTER OF HOPE

LOCATION: Westmingster, Colorado.
LATITUDE: 40 degrees N.
REGION: Temperate.
ARCHITECT: J. K. Abrams, Westmingster, Colorado.
SOLAR SYSTEM DESIGN, FABRICATION AND
    INSTALLATION: R. M. Products, Denver, Colorado.

SPACIAL SOLUTION

This building functions as a church and a day-care center, with floor areas of 2,700 square feet and 17,300 square feet, respectively. Its construction is wood frame, with the walls insulated to $R=17$ and the ceiling insulated to $R=31$. The main entrance has a vestibule to reduce energy losses. All windows are single glazed, with 50 percent of them facing south.

MECHANICAL SOLUTION

The church center maintains an uneven heating pattern, with the church area being unheated five to six days a week. When it *is* heated, the other areas are allowed to lower their average temperature.

There are 3,100 square feet of water-cooled, flat-plate collectors integrated into the day-care center roof at a tilt of 55 degrees from the horizontal. The system includes an automatic draining valve, which operates before the water can freeze. The heated water flows directly from the collector to the storage. The storage is composed of two tanks with capacities of 6,000 and 8,000 gallons. The 6,000-gallon storage tank functions in conjunction with a carrier heat pump.

The building requires no auxiliary or back-up element because the combination of solar collector, storage and heat pump provides 100 percent of the heating requirements.

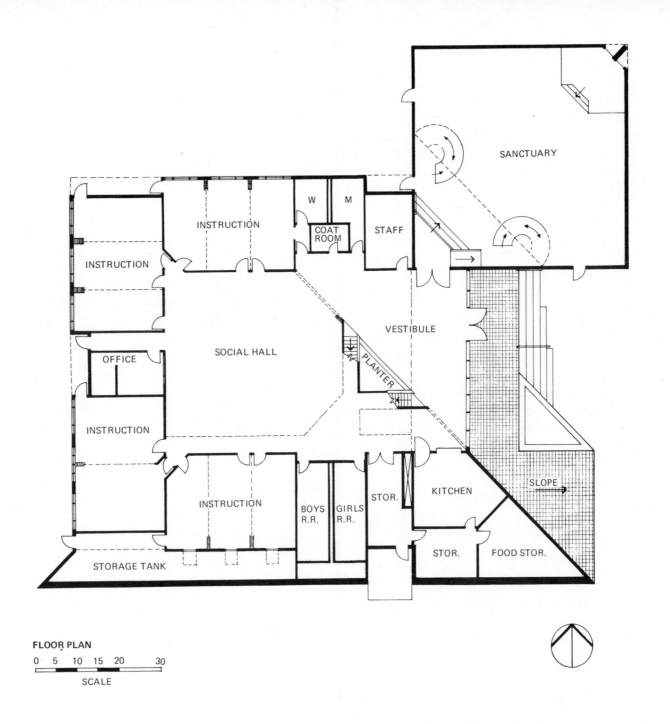

SANCTUARY

INSTRUCTION

INSTRUCTION

W   M

COAT ROOM

STAFF

OFFICE

INSTRUCTION

SOCIAL HALL

VESTIBULE

PLANTER

INSTRUCTION

BOYS R.R.

GIRLS R.R.

STOR.

KITCHEN

SLOPE

STORAGE TANK

STOR.

FOOD STOR.

FLOOR PLAN

0   5   10   15   20   30

SCALE

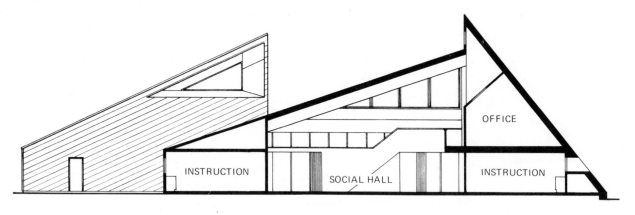

NORTH–SOUTH SECTION

SCALE

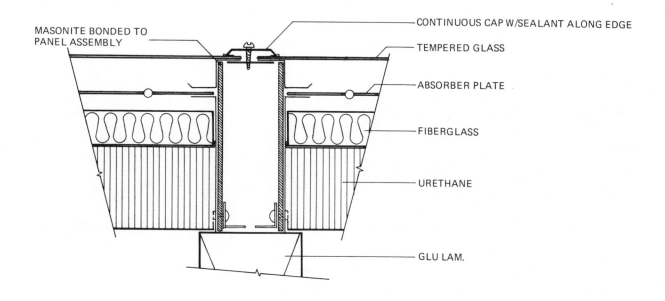

MOUNTING DETAIL

MASONITE BONDED TO
PANEL ASSEMBLY

CONTINUOUS CAP W/SEALANT ALONG EDGE

TEMPERED GLASS

ABSORBER PLATE

FIBERGLASS

URETHANE

GLU LAM.

OFFICE

INSTRUCTION

SOCIAL HALL

INSTRUCTION

33

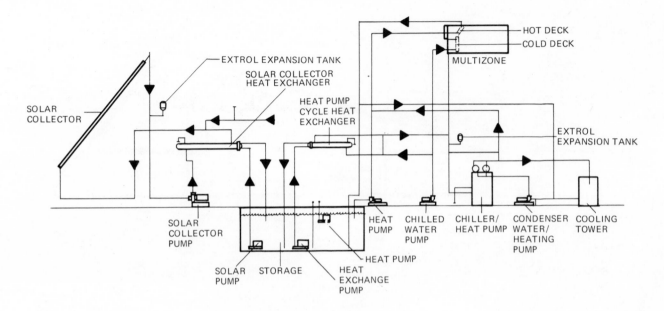

SOLAR SYSTEM SCHEMATIC

Labels in schematic:
- SOLAR COLLECTOR
- EXTROL EXPANSION TANK
- SOLAR COLLECTOR HEAT EXCHANGER
- HEAT PUMP CYCLE HEAT EXCHANGER
- HOT DECK
- COLD DECK
- MULTIZONE
- EXTROL EXPANSION TANK
- SOLAR COLLECTOR PUMP
- HEAT PUMP
- CHILLED WATER PUMP
- CHILLER/ HEAT PUMP
- CONDENSER WATER/ HEATING PUMP
- COOLING TOWER
- SOLAR PUMP
- STORAGE
- HEAT EXCHANGE PUMP
- HEAT PUMP

Main entrance—east elevation

Northwest elevation

Vestibule

South elevation

Southwest elevation

35

COMMUNITY COLLEGE OF DENVER, NORTH CAMPUS

LOCATION: Westmingster, Colorado.

LATITUDE: 40 degrees N.

REGION: Temperate.

ARCHITECT: A.B.R. Partnership, Denver, Colorado.

ENGINEER: Bridges & Paxton, Albuquerque, New Mexico.

SPACIAL SOLUTION

The Public Service of Colorado issued a bulletin in 1973 informing all architects that gas hook-ups for new construction could only be assured if applications were made by January 1, 1974. With this and future gas curtailment in mind, the architects for this college were forced to look for an alternative energy source. The fact that Denver enjoys the nation's highest solar radiation receptivity, coupled with the gas restrictions, pointed the way for a solar energized college.

The building has 300,000 square feet of floor area and serves 3,500 full time students. The wall construction is insulated block, with an $R=15$ value. The ceiling insulating value is $R=12$. All windows are double glazed and comprise 12 percent of the total exterior surface area.

MECHANICAL SOLUTION

Due to the square footage and the volume incorporated by the building, a combination of solar heat, heat recovery pumps, redistribution elements and several highly sophisticated controls were necessary to produce the desired tempered environment. There are 35,000 square feet of liquid-cooled, flat-plate collectors mounted on two parallel steel frames that are tilted to 55 degrees from the horizontal. The heat removal medium is a combination of water and 30 percent ethylene glycol, with the heat exchanger between the collector and the storage. The ratio of floor to collector area is approximately 10 to 1.

The storage is composed of two 100,000-gallon-capacity poured concrete tanks. They combine with the chiller-heat pump to deliver heat to the fan-coil system. Tank water temperatures may drop to 50 degrees without affecting the system's efficiency. When the water temperature exceeds 100 degrees F, the heat pump element is bypassed. In addition, the pumps act to redistribute heat from areas where there is a surplus to areas where it is needed. Heat recovery devices are located in the exhaust fans to extract what would otherwise be lost heat. The absorption chiller provides summer cooling.

The combination of all these systems provides 100 percent of the space heating, 80 percent of the domestic hot water and 100 percent of the space cooling requirements. The remaining 20 percent needed for the domestic hot water is supplied by a gas fired heater.

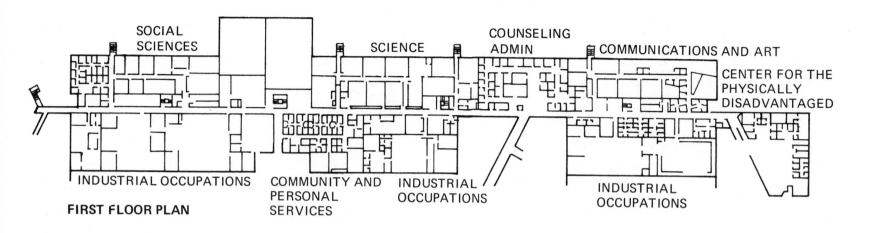

SOCIAL
SCIENCES

SCIENCE

COUNSELING
ADMIN

COMMUNICATIONS AND ART

CENTER FOR THE
PHYSICALLY
DISADVANTAGED

INDUSTRIAL OCCUPATIONS

COMMUNITY AND
PERSONAL
SERVICES

INDUSTRIAL
OCCUPATIONS

INDUSTRIAL
OCCUPATIONS

FIRST FLOOR PLAN

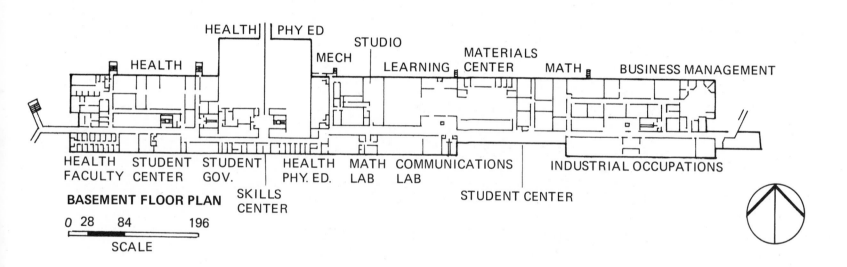

HEALTH

PHY ED

STUDIO

MECH

MATERIALS
CENTER

HEALTH

LEARNING

MATH

BUSINESS MANAGEMENT

HEALTH
FACULTY

STUDENT
CENTER

STUDENT
GOV.

HEALTH
PHY. ED.

MATH
LAB

COMMUNICATIONS
LAB

INDUSTRIAL OCCUPATIONS

STUDENT CENTER

BASEMENT FLOOR PLAN

SKILLS
CENTER

0 28   84   196

SCALE

37

SYSTEM NO. 1 ENERGY FLOW DIAGRAM SHOWING WINTER CYCLE (BELOW 50 F)
DAY TIME OPERATION (HEAT PUMP OPERATION WITH HEAT RECOVERY FROM
LIGHT, SUNLIT SIDE (SOLAR) AND PEOPLE.)

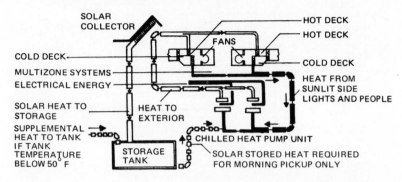

SOLAR
COLLECTOR

HOT DECK
HOT DECK

FANS

COLD DECK

COLD DECK

MULTIZONE SYSTEMS

ELECTRICAL ENERGY

HEAT FROM
SUNLIT SIDE
LIGHTS AND PEOPLE

SOLAR HEAT TO
STORAGE

HEAT TO
EXTERIOR

SUPPLEMENTAL
HEAT TO TANK
IF TANK
TEMPERATURE
BELOW 50° F

CHILLED HEAT PUMP UNIT

STORAGE
TANK

SOLAR STORED HEAT REQUIRED
FOR MORNING PICKUP ONLY

ENERGY FLOW DIAGRAM SHOWING WINTER CYCLE - NIGHTTIME OPERATION
HEAT PUMP OPERATION - SOLAR HEAT FROM STORAGE OR DIRECT SOLAR
HEATING FROM STORAGE TANK

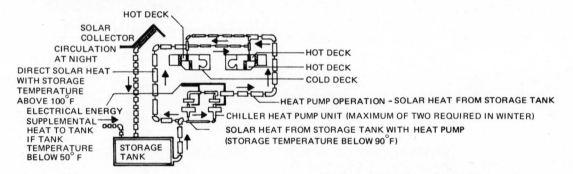

HOT DECK

SOLAR
COLLECTOR

CIRCULATION
AT NIGHT

HOT DECK
HOT DECK
COLD DECK

DIRECT SOLAR HEAT
WITH STORAGE
TEMPERATURE
ABOVE 100°F

ELECTRICAL ENERGY

HEAT PUMP OPERATION - SOLAR HEAT FROM STORAGE TANK

SUPPLEMENTAL
HEAT TO TANK
IF TANK
TEMPERATURE
BELOW 50° F

CHILLER HEAT PUMP UNIT (MAXIMUM OF TWO REQUIRED IN WINTER)

SOLAR HEAT FROM STORAGE TANK WITH HEAT PUMP
(STORAGE TEMPERATURE BELOW 90°F)

STORAGE
TANK

ENERGY FLOW DIAGRAM SHOWING COOLING CYCLE (OUTSIDE TEMPERATURE ABOVE 75°F)

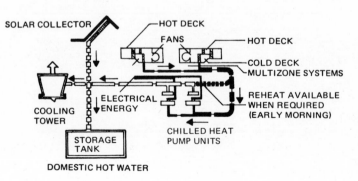

SOLAR COLLECTOR

HOT DECK

FANS

HOT DECK

COLD DECK
MULTIZONE SYSTEMS

ELECTRICAL
ENERGY

REHEAT AVAILABLE
WHEN REQUIRED
(EARLY MORNING)

COOLING
TOWER

CHILLED HEAT
PUMP UNITS

STORAGE
TANK

DOMESTIC HOT WATER

38

Main entrance—south elevation

Southeast elevation

Northwest elevation

Roof mounted collectors

Roof mounted equipment protected by collectors

Earth berms—Northeast corner

Circulation Tower—west elevation

40

Main level entrance—west elevation

ARMED FORCES RESERVE CENTER ARMORY

LOCATION: Norwich, Connecticut.
LATITUDE: 41.5 degrees N.
REGION: Temperate.
ARCHITECT: Moore, Grover, Harper, Hartford, Connecticut.
SOLAR CONSULTANT: A. D. Little, Massachusetts.

SPACIAL SOLUTION

This project involves two buildings: the armory, with 35,800 square feet of floor area; and the organizational maintenance shop, which has two stories, the first being used for storage space, and the second—with 7,500 square feet—used for offices. The offices and the shop are the only areas of the building kept at normal temperatures. The temperature of the storage space in the armory is allowed to drift, sometimes as low as 50 degrees F during the winter.

The windows comprise less than 10 percent of the exterior surface area, and the majority of them are located in the south elevation to aid the passive thermal qualities inherent in the walls and floor slab. All windows are double glazed and the ones facing south are shaded by the collectors.

MECHANICAL SOLUTION

The main achievement of this system is that it has reduced oil consumption to less than 13 percent, as compared with a conventional armory of similar size.

There are 3,640 square feet of liquid-cooled, flat-plate collectors, mounted on bolted frames and tilted to 60 degrees from the horizontal. The heat removal liquid is circulated through the panels to the heat exchanger located in the storage element.

41

The storage tank has a capacity of 2,000 gallons and serves the office area exclusively. The shop area makes use of the thermal storage capacity inherent in the walls and floor slab. The ratio of floor to collector area is approximately 3.5 to 1.

The solar system provides 70 percent of the armory's heating requirements; cooling is provided through conventional means. The domestic hot water is preheated by the collectors, and the required make-up energy is supplied by an oil fired boiler.

NORTH EAST ELEVATION

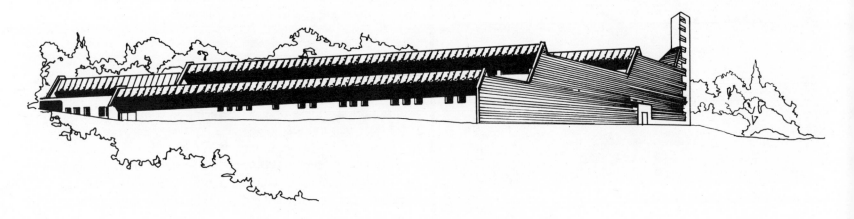

SOUTH EAST ELEVATION

UPTON MULTI-PURPOSE CENTER

LOCATION: Baltimore, Maryland.
LATITUDE: 39.2 degrees N.
REGION: Temperate.
ARCHITECT: William D. Potts, Baltimore, Maryland.
CONSULTING ENGINEER: Muller Associates, Baltimore,
   Maryland.

SPACIAL SOLUTION

This building is located in the Upton urban renewal area, which is immediately northwest of Baltimore's business district. The main design consideration was that the desired south orientation was at 45 degrees to the urban grid. The solution was to orient the core of the facility, a steel framed gymnasium, to the south. All other related spaces were constructed out of stuccoed block and were distributed off a skylighted corridor which surrounds three sides of the multi-purpose room.

The center has a total floor area of 22,000 square feet and houses recreational and day-care facilities. The recreation zone has spaces for crafts, clubs and game rooms. The day-care center accomodates 55 preschool children and 40 children after school. The architect made use of recessed double glazed windows and airlock entry to improve the building's thermal performance. The skylight received a reflective treatment to balance heat gain/loss against lighting levels.

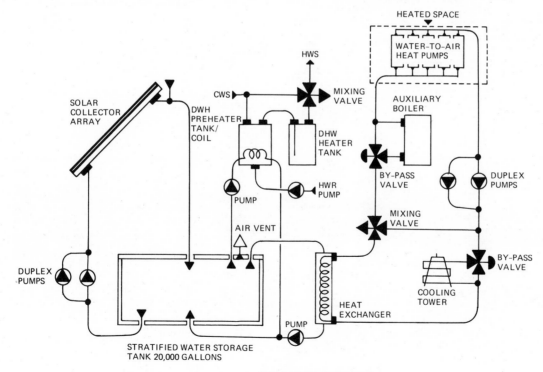

SOLAR SYSTEM SCHEMATIC

MECHANICAL SOLUTION

The solar system is used for space and water heating.

There are 3,480 square feet of water-cooled, flat-plate collectors, with a single cover plate, a copper absorbing plate and a non-selective surface. Freeze protection is provided through an automatic draining valve, which operates when the collector temperature is either less than 60 degrees, or 10 degrees less than the storage tank temperature, for a period of two hours.

The ratio of floor to collector area is approximately 6 to 1. The storage is a steel tank with a 20,000-gallon capacity. The tank feeds a liquid to liquid heat exchanger which, in turn, provides heat to a closed loop water to air heat pump. An electric boiler is used to maintain temperatures in the heat pump loop to the required minimum. The domestic water is preheated by circulating solar heated water from the storage tank to a heat exchanger in the domestic hot water tank. There is an energy recovery system which makes use of the exhausted air. The solar energy system supplies 71 percent of the annual heating needs.

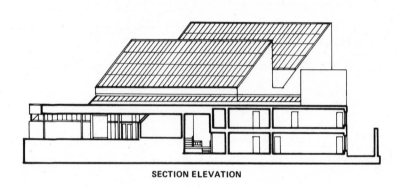

SECTION ELEVATION

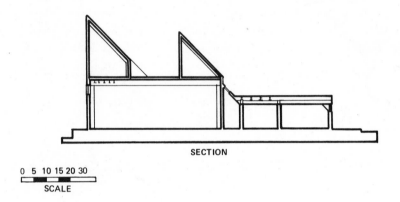

SECTION

0 5 10 15 20 30

SCALE

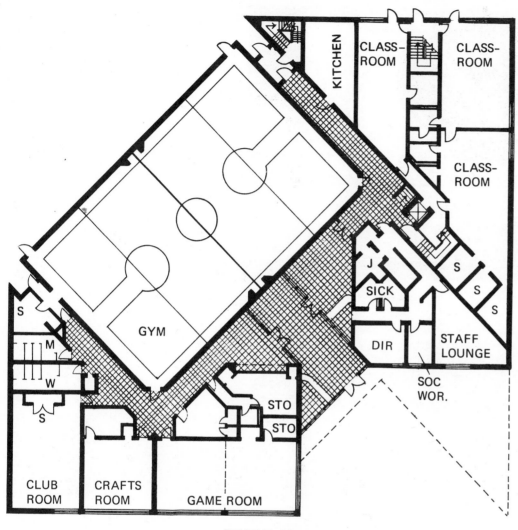
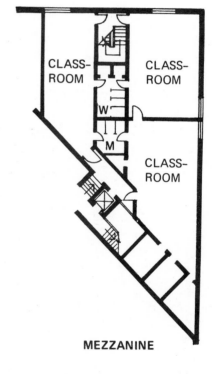

FIRST FLOOR PLAN

KITCHEN

CLASS-ROOM

CLASS-ROOM

CLASS-ROOM

GYM

S

J

SICK

S

S

S

DIR

STAFF LOUNGE

SOC WOR.

STO

STO

M

W

S

CLUB ROOM

CRAFTS ROOM

GAME ROOM

MEZZANINE

CLASS-ROOM

CLASS-ROOM

CLASS-ROOM

W

M

SCALE

0   5   10   20

45

Site view

South elevation

Photos by William D. Potts

46

PADONIA ELEMENTARY SCHOOL

LOCATION: Cockeysville, Maryland.
LATITUDE: 39.9 degrees N.
REGION: Temperate.
SOLAR SYSTEM DESIGN, MANUFACTURE AND
    INSTALLATION: AAI Corporation, Baltimore, Maryland.

SPACIAL SOLUTION

This project required the design, manufacture, retrofit installation and testing of the solar concentrating collector for the one-story, 45,000-square-foot elementary school. However, only 10,000 square feet are conditioned by the solar system. This area includes a cafetorium, library and administrative suite. Since this is a retrofit project, it does not use spacial energy saving concepts and limits the conservation efforts to the operation of the solar system.

MECHANICAL SOLUTION

The solar system was designed to provide space heating and cooling. The solar concentrating collector consists of a fixed parabolic reflector mirror, mounted on the school's south end. The mirror concentrates the solar radiation onto a liquid-cooled collector. The reflecting area is 3,388 square feet, and the solar collector area is 423 square feet. The ratio of reflector to collector area is 8 to 1, and the ratio of floor to reflector area is 2.9 to 1.

The storage is composed of two tanks, each with a 10,000-gallon capacity. One tank is used for hot water and the other for chilled water. The heated liquid from the collector is circulated to the heat exchanger in the tank, and from there the heated water flows to the fan coil units in the forced air system. An absorption chiller unit provides summer space cooling.

The solar mechanism provides 75 percent of the space heating and cooling demands. The domestic hot water is supplied through a conventional system which serves as the back-up.

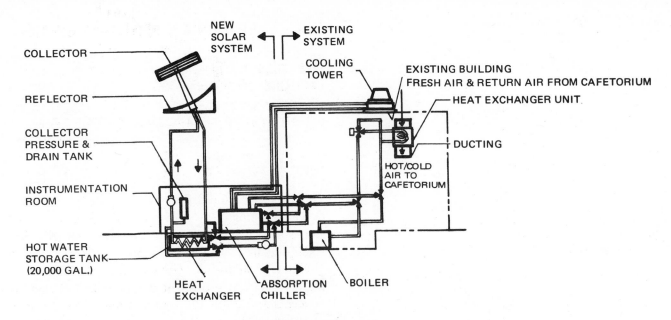

COLLECTOR

REFLECTOR

COLLECTOR
PRESSURE &
DRAIN TANK

INSTRUMENTATION
ROOM

HOT WATER
STORAGE TANK
(20,000 GAL.)

NEW
SOLAR
SYSTEM

EXISTING
SYSTEM

COOLING
TOWER

EXISTING BUILDING
FRESH AIR & RETURN AIR FROM CAFETORIUM

HEAT EXCHANGER UNIT

DUCTING

HOT/COLD
AIR TO
CAFETORIUM

HEAT
EXCHANGER

ABSORPTION
CHILLER

BOILER

SOLAR SYSTEM SCHEMATIC

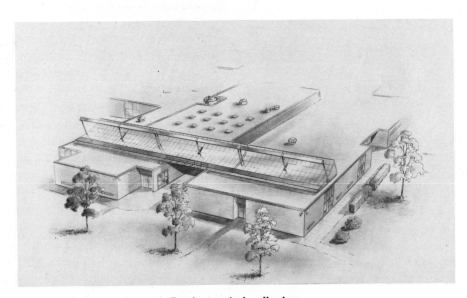

Roof mounted collectors

MASSACHUSETTS AUDUBON SOCIETY NATURE CENTER

LOCATION: Lincoln, Massachusetts.
LATITUDE: 42 degrees N.
REGION: Temperate.
ARCHITECT: Mass Design, Boston, Massachusetts.
SOLAR CONSULTANT: Solar Heat Corporation, Boston, Massachusetts.

SPACIAL SOLUTION

The Nature Center is housed in a 1951, two-volume, wood frame structure. The main wing, 70 feet by 40 feet, is used for educational purposes. The secondary wing, which is the newer addition, is 48 feet by 30 feet, and houses the gift shop.

The walls in the gift shop are insulated with 3-1/2 inches of formaldehyde foam; the roof has a resistance value of 19. The shop's foundation is insulated with 2 inches of styrofoam board. All windows are double glazed and the doors use an air-lock design.

MECHANICAL SOLUTION

This is a retrofit project. Alterations to the shop's roof were necessary to integrate the collector with the building's envelope. There are 960 square feet of water-cooled, flat-plate collectors mounted directly on the roof's plywood sheading, tilted to 45 degrees from the horizontal and oriented 15 degrees east of south. The system drains automatically to prevent the water from freezing. The heated water flows directly to the storage tank. The storage, which has a 2,000-gallon capacity, is cylindrical and placed in a vertical position.

The shop is heated by a baseboard radiator; the Nature Center is heated by convectors. The existing oil furnace provides the necessary back-up heat. The domestic hot water is preheated by a heat exchanger located in the storage tank.

The solar system provides 50 percent of the space heating and 75 percent of the water heating loads.

Northwest elevation

Southeast elevation

Southwest elevation

49

CAMBRIDGE SCHOOL

LOCATION: Weston, Massachusetts.
LATITUDE: 42 degrees, 22′ N.
REGION: Temperate.
ARCHITECT: Davies, Wolf and Bibbings, Cambridge, Massachusetts.
SOLAR ENGINEER: N. B. Saunders, Boston, Massachusetts.

SPACIAL SOLUTION

This building's west end is new, and the east end was remodeled after a fire. The basement floor is poured concrete. The exterior as well as the interior masonry walls are 12 inches thick. The exterior walls are insulated with 4 inches of fiberglass and are protected with a stucco layer. The building is rectangular in shape, with the longer sides facing north and south; the exact orientation is 26 degrees west of south. The total floor area is 14,400 square feet and is divided into classrooms, offices, baths, kitchen, dining room and the various other supportive functions needed in a school. Forty percent of the windows are double glazed and are located on the south wall to allow for passive heat gain. The remaining windows are single glazed. The roof contains a "staircase" skylight which allows solar radiation to penetrate and be stored in the massive walls and floor. In addition to supplying solar penetration for the passive system, the skylight allows a great deal of light to infiltrate the interior spaces, thereby effectively reducing the need for electric lighting. In the summer, the internal circulation of cool night air effectively reduces the wall temperature, resulting in cooler daytime spaces. The passive solar system provides 25 percent of the required heat. The remaining energy is supplied by an oil fired, hot water furnace.

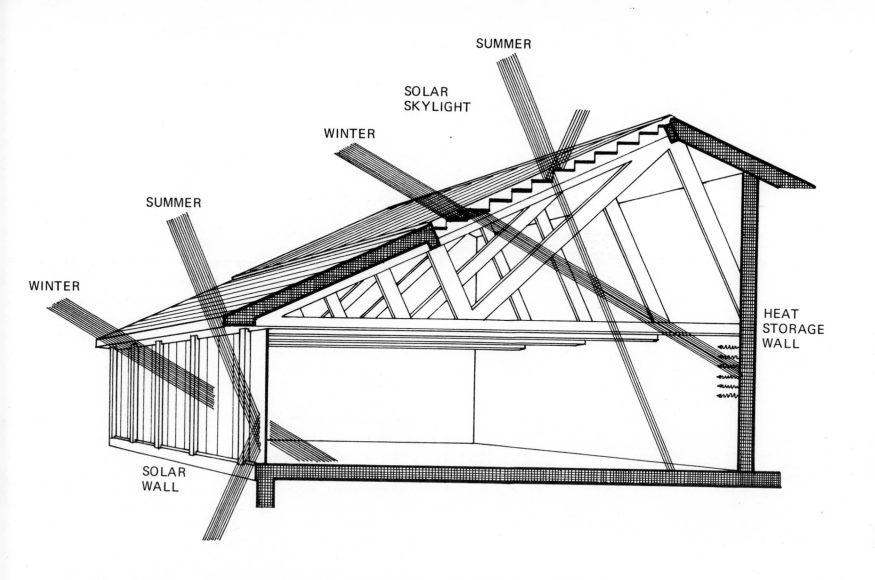

SUMMER

SOLAR
SKYLIGHT

WINTER

SUMMER

WINTER

HEAT
STORAGE
WALL

SOLAR
WALL

SOLAR SKYLIGHT, HEAT STORAGE WALL AND SOLAR WALL

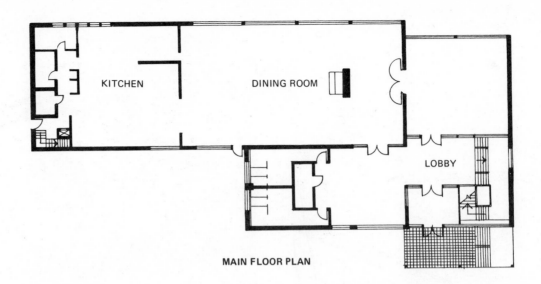

MAIN FLOOR PLAN

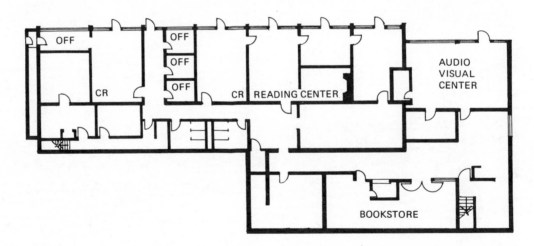

LOWER FLOOR PLAN

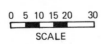

0   5 10 15 20   30

SCALE

Main entrance northwest elevation

Southeast elevation

Staircase skylight

53

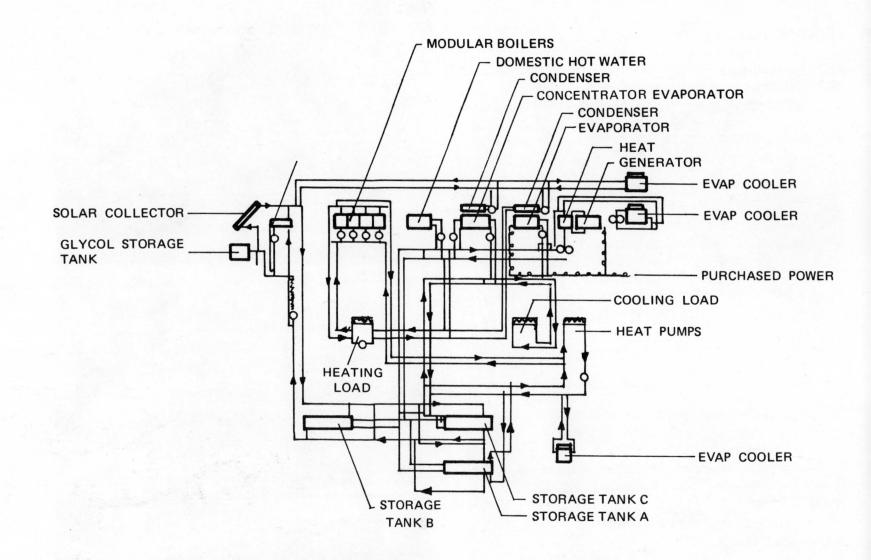

MODULAR BOILERS
DOMESTIC HOT WATER
CONDENSER
CONCENTRATOR EVAPORATOR
CONDENSER
EVAPORATOR
HEAT GENERATOR
EVAP COOLER
EVAP COOLER
SOLAR COLLECTOR
GLYCOL STORAGE TANK
PURCHASED POWER
COOLING LOAD
HEAT PUMPS
HEATING LOAD
EVAP COOLER
STORAGE TANK C
STORAGE TANK A
STORAGE TANK B

SOLAR SYSTEM SCHEMATIC

54

NORRIS COTTON FEDERAL BUILDING

LOCATION: Manchester, New Hampshire.
LATITUDE: 43 degrees N.
REGION: Temperate.
ARCHITECT: Issack & Issack, New York, New York.
SOLAR ENGINEER: Dubin Bloome Associates. New York, New York.
INSTRUMENTATION, DESIGN AND OPERATION: National Bureau of Standards.

SPACIAL SOLUTION

This building has 57,650 square feet of office space, used by over 400 employees of six federal agencies and distributed among seven floors. In addition, it provides 71,800 square feet of parking and 6,900 square feet for a mechanical penthouse. The envelope has been shaped into a cubic form, with no windows on the north side and with fixed shading controls on the east, west and south elevations. The controls allow winter light into the building but shade direct summer light. This is done with 85 percent-effectiveness. In general, the windows are smaller than those used in similar conventional buildings. Also, they have venetian blinds between the two glass layers to allow the users to control the penetration of natural light.

Different lighting systems are used for each floor. This is one of several ongoing experiments designed to test and improve technical construction aspects. Another experiment was to introduce positive pressure (higher than outside air pressure) to prevent infiltration.

MECHANICAL SOLUTION

The solar system is used for space heating and cooling, as well as for domestic water heating.

There are 4,600 square feet of collectors mounted in four rows, with manual tilt mechanisms which slope from 20 to 80

Northwest elevation

Sun protecting screens—east elevation

55

South elevation

degrees from the horizontal. Although each collector array was purchased from a different manufacturer, all use a heat removal medium of water and ethylene glycol, circulated to the heat exchangers by a 10-horsepower centrifugal pump.

There are three cylindrical steel tanks located in the basement, each with a 10,000-gallon capacity.

Each floor is heated by a different means: the first, second and third floors use solar assisted heat pumps; the fourth floor uses baseboard radiators heated with water from the storage tank; and the fifth, sixth and seventh floors use a forced air system, which is also assisted by the storage tanks. A combination back-up system is used, including a gas fired boiler, heat pumps and heat recovery cycles.

The solar system provides 30 percent of the total energy demand. However, due to the highly efficient energy conscious design, the building consumes 50 percent less energy than other comparable buildings.

TROY MIAMI COUNTY LIBRARY

LOCATION: Troy, Ohio.
LATITUDE: 40 degrees N.
REGION: Temperate.
SOLAR ENGINEER: Research Institute, University of Dayton.
SOLAR EQUIPMENT MANUFACTURER: Ownes Illinois
   Company, Chicago, Illinois.

SPACIAL SOLUTION

The public library is located in the downtown area of the city. Its original design called for revisions to be made for the eventual retrofit to solar energy use. This procedure is consistent with the city's expressed interest in becoming a model solar city. Plans have been developed to provide financial support to design solar energy systems for every new building constructed in the city. The library has a total floor area of 22,600 square feet, of which 16,000 square feet are contained in the first floor, with the rest located in the basement. Wall construction is of 4-inch brick with 8-inch block and 3 inches of insulation. This yields a "U" value of 0.09. The roof has 6 inches of fiberglass insulation and its "U" value is 0.04.

There are 1,800 square feet of double glazed recessed windows. This is less than 20 percent of the over-all surface area.

MECHANICAL SOLUTION

The retrofit system was designed for space heating.

There are 3,600 square feet of triple tube, vacuum-jacketed, water-cooled tube assembled collectors, tilted to 23.5 degrees from the horizontal. The heat loss from the tubes is so small that even during cold winter nights, no freezing occurs. Because of this, the heat removing agent is 100 percent water. In special cases, when the exterior temperatures are extremely low, warm water from the storage is circulated for a few minutes through the tubes. Some of the existing trees were trimmed to permit maximum solar penetration to the collector area.

The cylindrical fiberglass tank has a 1,000-gallon capacity and is located below grade on the west end. Hot water is circulated from the tank to fan coil units in the forced air system. If needed, the tank is bypassed and water flows directly from the collector to the fan coils. The solar system provides 75 percent of the needed heat. The make-up energy is supplied by electric heaters in the forced air system. Cooling is provided by a conventionally powered water chiller, designed for eventual integration into the solar system.

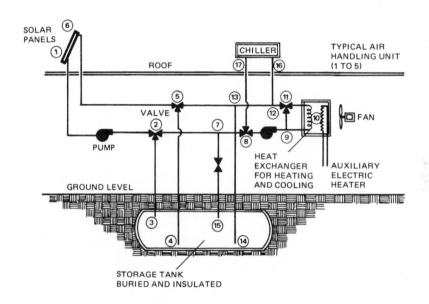

TYPICAL HEATING AND COOLING CYCLES	
SOLAR HEAT TO STORAGE	①-②-③-④-⑤-⑥-①
DIRECT SOLAR HEATING	①-②-⑦-⑧-⑨-⑩-⑪-⑫-⑬-⑤-⑥-①
HEATING FROM STORAGE	⑮-⑦-⑧-⑨-⑩-⑪-⑫-⑬-⑭-⑮
AUXILIARY HEATING	ELECTRIC RESISTANCE HEATING
CONVENTIONAL COOLING	⑰-⑧-⑨-⑩-⑪-⑫-⑯-⑰

SOLAR SYSTEM SCHEMATIC

RIDLEY PARK POST OFFICE

LOCATION: Ridley Park, Pennsylvania.
LATITUDE: 40 degrees N.
REGION: Temperate.
ARCHITECT: Environmental Design Collaborative, Philadelphia, Pennsylvania.
SOLAR CONSULTANT: Honeywell Inc., Illinois.
MECHANICAL ENGINEER: Cooley & Erickson, Philadelphia, Pennsylvania.

SPACIAL SOLUTION

This post office has 6,250 square feet of floor area on one level. It uses insulated porcelain forms with a continuous exterior skin and nonoperablewindows, all in an effort to reduce heating and cooling loads.

MECHANICAL SOLUTION

This is the first new post office building in the United States to be heated and cooled with solar energy.

There are 2,500 square feet of liquid-cooled, flat-plate collectors mounted on steel frames on the flat roof and tilted to 30 degrees from the horizontal. They were produced by Chamberlain Manufacturing Corp. The heat removal liquid circulates from the collector to the 4,000-gallon capacity storage tank, which is located above grade. Heat is distributed to the rooms by means of fan-coil units located in the forced air system. During the summer, the collector provides water at 190 degrees F to operate the 25-ton water fired absorption cooling mechanism.

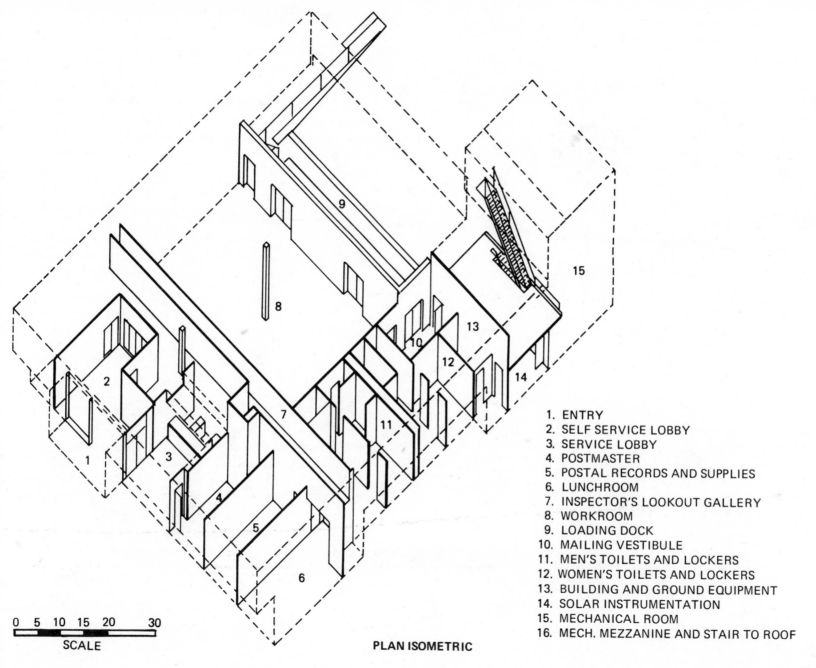

1. ENTRY
2. SELF SERVICE LOBBY
3. SERVICE LOBBY
4. POSTMASTER
5. POSTAL RECORDS AND SUPPLIES
6. LUNCHROOM
7. INSPECTOR'S LOOKOUT GALLERY
8. WORKROOM
9. LOADING DOCK
10. MAILING VESTIBULE
11. MEN'S TOILETS AND LOCKERS
12. WOMEN'S TOILETS AND LOCKERS
13. BUILDING AND GROUND EQUIPMENT
14. SOLAR INSTRUMENTATION
15. MECHANICAL ROOM
16. MECH. MEZZANINE AND STAIR TO ROOF

0 5 10 15 20 30
SCALE

PLAN ISOMETRIC

59

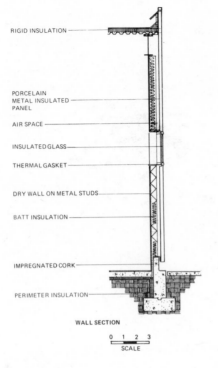

RIGID INSULATION

PORCELAIN
METAL INSULATED
PANEL

AIR SPACE

INSULATED GLASS

THERMAL GASKET

DRY WALL ON METAL STUDS

BATT INSULATION

IMPREGNATED CORK

PERIMETER INSULATION

WALL SECTION

0 1 2 3
SCALE

1. SUPPLY PIPING
2. RETURN PIPING
3. HEAT EXCHANGERS
4. PUMPS
5. ABSORPTION WATER UNIT
6. STORAGE TANK
7. BOILER
8. AIR HANDLING UNIT
9. FINNED TUBE RADIATION
10. COOLING TOWER
11. FLUID COOLER
12. COLLECTOR ARRAY

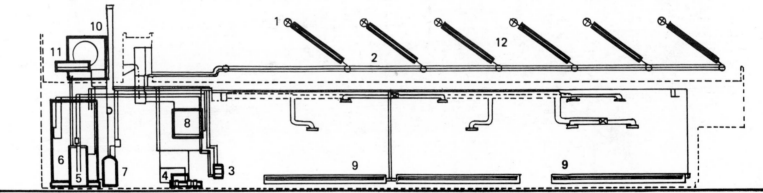

MECHANICAL SECTION

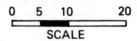

0 5 10 20
SCALE

South view

TERRASET ELEMENTARY SCHOOL

LOCATION: Reston, Virginia.
LATITUDE: 39 degrees N.
REGION: Temperate.
ARCHITECT: Davis, Smith, Carter, Arlington, Virginia.
ENGINEER: Vizant Associates, Arlington, Virginia.
SOLAR CONSULTANT: Smith, Hichman & Grylls Associates,
    Detroit, Michigan.

SPACIAL SOLUTION

The design requirements were to construct a totally energy conscious school to accomodate 1,000 students in a wooded 14-acre site. The school was integrated with the surrounding topography by berming it with 3 feet of earth above and around the roof and walls. The shell is built of reinforced concrete, and the interior partitions are graphically painted block. This heavy construction creates a thermal mass which couples with the earth's insulation to produce an environment that has small heating needs.

The plan is composed of four circular pods, used as teaching areas, grouped around the media center. Offices and supportive facilities are located to the north. The total area of 65,000 square feet was reduced in volume, wherever possible, by lowering the ceiling height from 10 feet to 8 feet. Windows are single glazed and compose 20 percent of the gross wall area. They are protected from the summer sun by deep overhangs.

MECHANICAL SOLUTION

The original budget did not make provisions for a solar system. The architects applied to various federal agencies for partial funding of the system, but did not receive an answer. However, they were approached by a representative of the Saudi Arabian Government with a large grant for the design and installation of a solar heating and cooling system.

There are 7,000 square feet of tubular glass, water-cooled, vacuum-jacketed collectors mounted on an elevated steel space frame. The collection system is complemented by a double bundle heat exchanger and variable volume heat redistribution system. The only heat required during the normal school year is that used to offset daily perimeter losses near the windows and the small amount used to maintain minimum temperature levels at night.

There are three 10,000-gallon capacity water storage tanks, each capable of storing hot or chilled water.

Space cooling is required at all times. This is done by a 50-ton absorption chiller unit, powered by hot water from the solar system. A cooling tower is needed to dispense the heat from the chillers. Back-up energy is rendered by a reciprocating chiller and electric heaters. The school has an annual energy savings of 75 percent as compared to conventional schools of similar size.

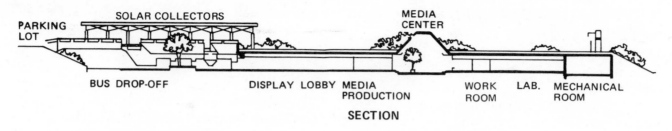

PARKING LOT   SOLAR COLLECTORS   MEDIA CENTER

BUS DROP-OFF   DISPLAY LOBBY MEDIA PRODUCTION   WORK ROOM   LAB.   MECHANICAL ROOM

SECTION

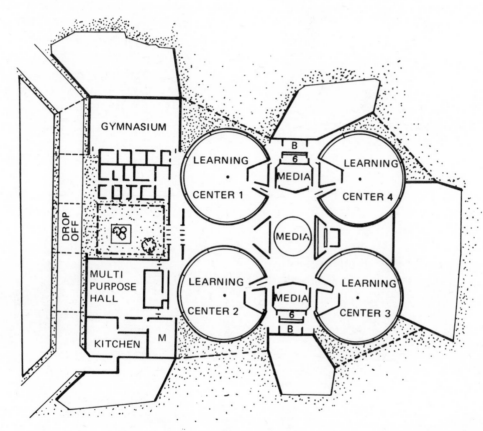

GYMNASIUM

DROP OFF

MULTI PURPOSE HALL

KITCHEN   M

LEARNING CENTER 1

LEARNING CENTER 2

LEARNING CENTER 4

LEARNING CENTER 3

B
6
MEDIA

MEDIA

MEDIA
6
B

FLOOR PLAN

Northwest elevation

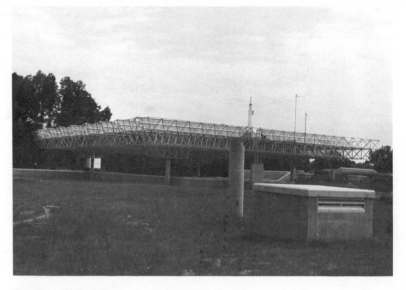

Southwest elevation

Collectors mounted on space frame

Mechanical vents north elevation

Detail of mechanical vents

FAQUIER COUNTY PUBLIC HIGH SCHOOL

LOCATION: Warrenton, Virginia.
LATITUDE: 39 degrees N.
REGION: Temperate.
SOLAR SYSTEM DESIGN AND
    INSTALLATION: Intertechnology Corporation, Fairfax,
Virginia.

MECHANICAL SOLUTION

Intertechnology Corporation designed and installed a solar heating system for this set of five separate classroom buildings with a combined floor area of 4,100 square feet. There are 2,500 square feet of water-cooled, flat-plate collectors mounted on a scaffold and tilted to 53 degrees from the horizontal. No anti-freeze is used; however, corrosion inhibitors are utilized. An automatic draining valve can empty the panels in 1-1/2 minutes. Two reinforced concrete tanks with a total capacity of 11,000 gallons are located underground and are heavily insulated. This retrofit system provides 60 percent of the space heating needs, with the back-up energy being supplied by an existing oil fired furnace and electric heaters.

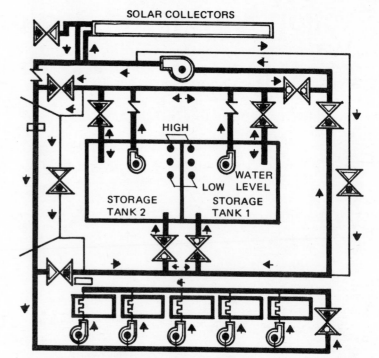

SOLAR SYSTEM SCHEMATIC

South facing collectors

Section 3. Hot Arid Region

The buildings presented in this section are the following:

Green Elementary School. San Diego, California.

New Mexico Department of Agriculture. Las Cruces, New Mexico.

Nambe Indian Community Center. Nambe, New Mexico.

The Benedictine Monastery. Pecos, New Mexico.

National Security and Resources Study Center. Los Alamos, New Mexico.

Tempe Union High School. Tempe, Arizona.

GREEN ELEMENTARY SCHOOL

LOCATION: San Diego, California.

LATITUDE: 32 degrees N.

REGION: Hot Arid.

ARCHITECT: Robert Des Lauries, La Mesa, California.

ENGINEER: Hugh Carter Engineering Corporation, San Francisco, California.

SPACIAL SOLUTION

When officials in the San Diego School District decided to build ten new schools in 1974, they looked into the possibility of using solar heat. The Green Elementary School was chosen as a testing ground to compare two identical 10,000-square-foot buildings; one heated by the sun and the other by conventional means. Both were equipped with the necessary solar hook-ups to expedite the eventual retrofitting in case the experiment proved successful.

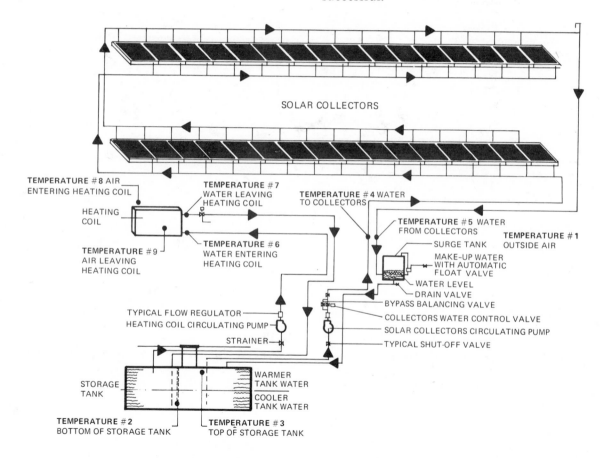

SOLAR COLLECTORS

TEMPERATURE #8 AIR ENTERING HEATING COIL

TEMPERATURE #7 WATER LEAVING HEATING COIL

TEMPERATURE #4 WATER TO COLLECTORS

HEATING COIL

TEMPERATURE #5 WATER FROM COLLECTORS

TEMPERATURE #1 OUTSIDE AIR

SURGE TANK

TEMPERATURE #9 AIR LEAVING HEATING COIL

TEMPERATURE #6 WATER ENTERING HEATING COIL

MAKE-UP WATER WITH AUTOMATIC FLOAT VALVE

WATER LEVEL

DRAIN VALVE

BYPASS BALANCING VALVE

TYPICAL FLOW REGULATOR

HEATING COIL CIRCULATING PUMP

COLLECTORS WATER CONTROL VALVE

SOLAR COLLECTORS CIRCULATING PUMP

STRAINER

TYPICAL SHUT-OFF VALVE

STORAGE TANK

WARMER TANK WATER

COOLER TANK WATER

TEMPERATURE #2 BOTTOM OF STORAGE TANK

TEMPERATURE #3 TOP OF STORAGE TANK

SOLAR SYSTEM DIAGRAM

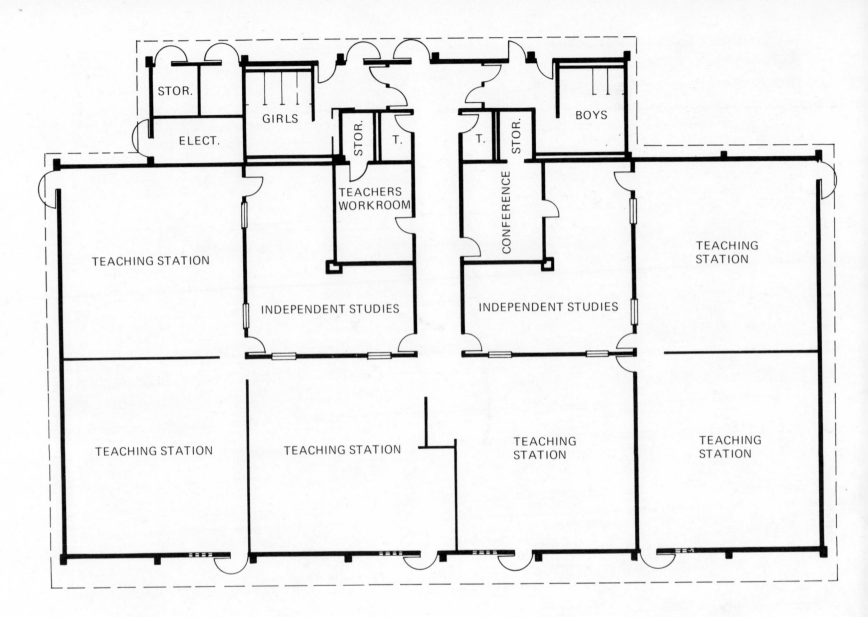

STOR.

ELECT.

GIRLS

STOR.

T.

TEACHERS WORKROOM

CONFERENCE STOR.

T.

BOYS

TEACHING STATION

INDEPENDENT STUDIES

INDEPENDENT STUDIES

TEACHING STATION

TEACHING STATION

TEACHING STATION

TEACHING STATION

TEACHING STATION

TEACHING STATION

PLAN

0   5   10   15   20   25 FEET

SCALE

Center court yard

Main entrance

Photo by George Lyons

The school's exterior walls are made of concrete, and the roof system rests on pre-cast, "T"-shaped, concrete beams. All windows face south and comprise less than 2 percent of the total exterior surface area. The main entrance is from the south and it uses a vestibule to reduce heat losses.

MECHANICAL SOLUTION

The solar system was designed for space heating. It uses 660 square feet of flat-plate, liquid-cooled collectors, mounted on the flat roof in two arrays and tilted to 42 degrees from the horizontal. Water is circulated through the collectors until the storage tank temperature reaches 140 degrees F, at which time the collectors are automatically drained. This also takes place at night to prevent heat losses. The heated water is circulated from the 5,000-gallon storage tank to the fan-coils in the air handling unit, which is also equipped with electric heating coils for back-up. On the average, the solar heated building is 5,800 Kilowatt hours more efficient than its counterpart.

South elevation

Roof mounted collectors

74

NEW MEXICO DEPARTMENT OF AGRICULTURE

LOCATION: Las Cruces, New Mexico.
LATITUDE: 32.3 degrees N.
REGION: Hot Arid.
ARCHITECT: W. T. Harris and Associates, Las Cruses, New Mexico.
MECHANICAL ENGINEER: Bridges & Paxton, Albuquerque, New Mexico.

SPACIAL SOLUTION

This building houses offices and laboratories for the study of crop diseases and pesticides. Its rectangular plane is 130 feet by 177 feet, with the main floor measuring 23,385 square feet and the basement and mechanical rooms measuring 2,149 square feet. The exterior walls are made of 12-inch block, filled with loose vermiculate insulation, with 3/4-inch stucco covering the exterior surface. In addition, a 1-inch urathane board with 3/4-inch sheet rock is used for the interior space surface. The ceiling

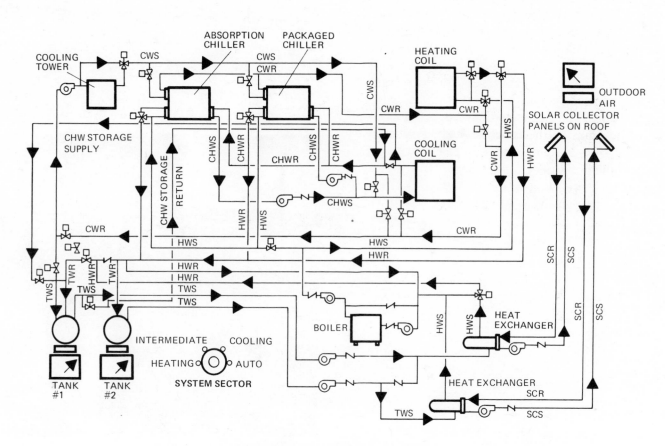

SOLAR HEATING AND COOLING GRAPHIC PANEL

Southeast elevation

has 6 inches of batt insulation plus 1 inch of rigid insulation. Alı
windows are double glazed and their area comprises less than 5
percent of the exterior surface.

MECHANICAL SOLUTION

The solar system provides space heating and cooling and domestic hot water, its major achievement is the capacity to heat and
cool during the same day.

There are 6,730 square feet of flat-plate, liquid-cooled collectors and 540 square feet of concentrating, tracking collectors,
both mounted to steel frames on the flat roof and tilted to 30
degrees from the horizontal. The heat removal medium is a combination of water and ethylene glycol, which flows through both
collector types. The heated liquid flows to the heat exchangers
located outside the two 15,000-gallon storage tanks. The heated
water flows from the tanks to the fan-coil forced air system.
Cooling is provided by a 100-ton and a 25-ton absorption
chiller, powered by the hot water in the storage.

The combination of these systems provides over 97 percent of
the space heating, 60 percent of the space cooling and 100 percent of the domestic water heating.

Roof mounted collectors

Experimental tracking collectors

Southeast elevation

NAMBE INDIAN COMMUNITY CENTER

LOCATION: Nambe, New Mexico.
LATITUDE: 35.9 degrees N.
REGION: Hot Arid.
ARCHITECT: A. L. McNown, Albuquerque, New Mexico.
SOLAR ENGINEER: Los Alamos Scientific Laboratory of the
    University of California.

SPACIAL SOLUTION

This one-story community center houses two meeting rooms, tribal offices, classrooms and a library. The architecture is of the southwest pueblo style and the construction is of traditional adobe. The exterior wall consists of 14 inches of adobe, 2 inches of polyurethane foam and a layer of stucco. The interior walls have 8-inch pume block, and the floor is a concrete slab 4 inches thick. All windows are double glazed and deeply recessed.

MECHANICAL SOLUTION

Since the operation of the active solar system has been integrated with that of the passive solar system, the over-all building performance is explained under one heading.

There are 440 square feet of air-cooled, flat-plate collectors integrated into the southwest wall of the mechanical penthouse. The collector is 7 feet high and 70 feet long, with a highly reflective sheet of anodized aluminum at its base and extending 18 feet to the south. This increases the total solar radiation received by the collector by 40 percent. The storage is 576 cubic feet in volume and contains 20 tons of stone. Air from the collector is circulated down through the bin. An air handling unit delivers the heated air from the storage to the rooms. The building's large thermal storage capacity provides a considerable percentage of its heating needs. For cooling, night air is circulated through the bin, effectively cooling the rocks (although little cooling is needed, due to the excellent use of insulation). The

combination of passive and active systems provides 65 percent of the space heating, with the remainder being supplied by a gas fired furnace.

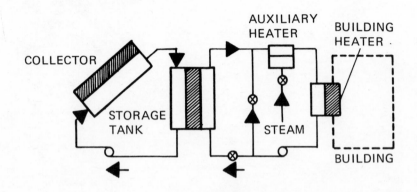

SOLAR SYSTEM SCHEMATIC

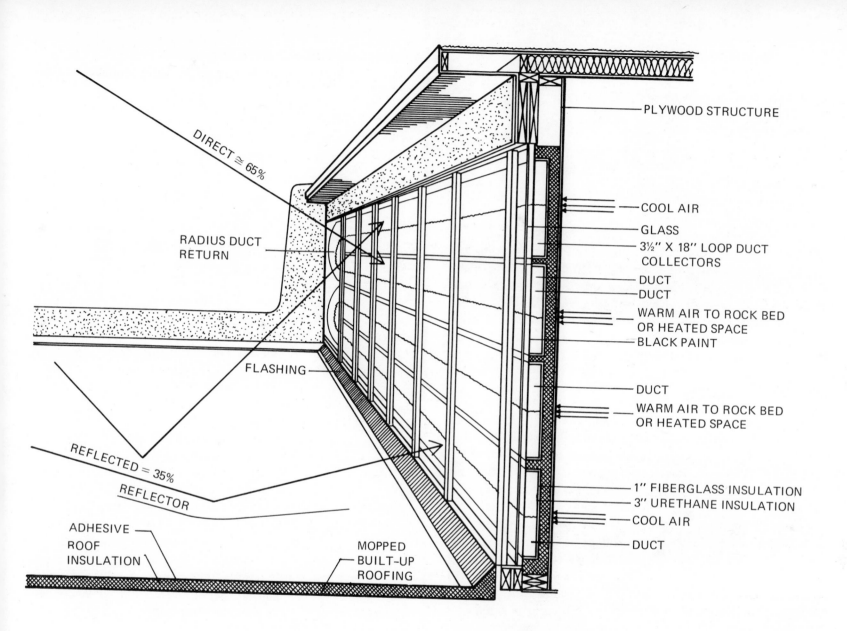

DIRECT ≅ 65%

RADIUS DUCT
RETURN

REFLECTED = 35%

REFLECTOR

FLASHING

ADHESIVE
ROOF
INSULATION

MOPPED
BUILT-UP
ROOFING

PLYWOOD STRUCTURE

COOL AIR

GLASS

3½" X 18" LOOP DUCT
COLLECTORS

DUCT
DUCT

WARM AIR TO ROCK BED
OR HEATED SPACE

BLACK PAINT

DUCT

WARM AIR TO ROCK BED
OR HEATED SPACE

1" FIBERGLASS INSULATION
3" URETHANE INSULATION

COOL AIR

DUCT

SECTION THRU COLLECTOR AND REFLECTOR

79

Southwest elevation

Southeast elevation

THE BENEDICTINE MONASTERY

LOCATION: Pecos, New Mexico.
LATITUDE: 35.5 degrees N.
REGION: Hot Arid.
ARCHITECT: Mike Hanse, Albuquerque, New Mexico.
SOLAR ENGINEER: Zomeworks Corporation, Albuquerque,
    New Mexico.

SPACIAL SOLUTION

When the monastery experienced a drastic increase in its publishing activities, a demand was created for a large warehouse in which to store printed material prior to distribution.

The passive design for the building was closely adapted to its function. The monastery has 10,000 square feet of floor area, including an unheated basement. The south portion of the building contains the mail room and offices, as well as living quarters for one full time occupant. The north portion is used as the warehouse. The outer walls are concrete block, with 2 inches of styrofoam insulation. The foundation also has styrofoam insulation around it. In the interior, a central block wall, filled with sand, separates the offices from the warehouse and provides additional thermal mass. The main feature of the passive system is the south-facing, water-filled drumwall, designed for thermal collection and storage. It consists of 138 55-gallon barrels, painted black and filled with propylene glycol and water, with 8 ounces of corrosion inhibitors and an equal amount of motor oil. The drums are 90 percent full to allow for thermal expansion. The offices are heated by a combination of direct solar gain and heat released from the drum wall. The warehouse, which is maintained at a lower temperature than that of the offices, is heated by direct heat gain through the clerestory. Vents connecting the offices and the warehouse may be opened to circulate the excess heat into the storage area.

Radiant electric heaters are used for back-up heat, and a 40-square-foot, water-cooled, flat-plate collector is used to provide domestic hot water.

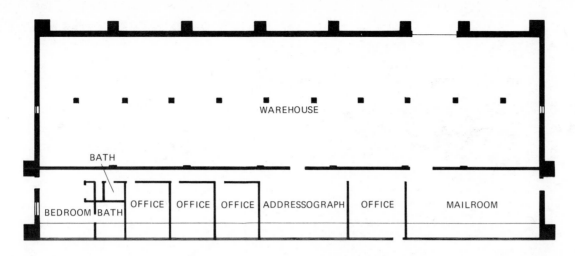

WAREHOUSE

BATH

BEDROOM BATH OFFICE OFFICE OFFICE ADDRESSOGRAPH OFFICE MAILROOM

WAREHOUSE FLOOR PLAN

0 5 10 20
SCALE

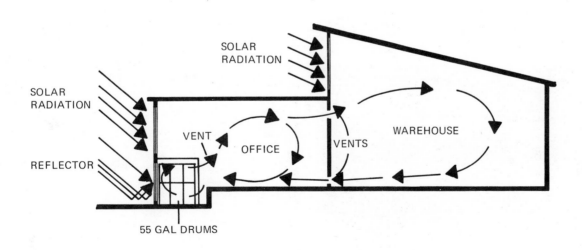

SOLAR
RADIATION

SOLAR
RADIATION

SOLAR
RADIATION

REFLECTOR

VENT

OFFICE

VENTS

WAREHOUSE

55 GAL DRUMS

SOLAR SYSTEM SCHEMATIC

81

Southeast elevation

Southwest elevation

NATIONAL SECURITY AND RESOURCES STUDY CENTER

LOCATION: Los Alamos, New Mexico.
LATITUDE: 35 degrees N.
REGION: Hot Arid.
ARCHITECT: S. Burnett, Tucson, New Mexico.
SOLAR CONSULTANT: J. Weingart, Tucson, New Mexico.

MECHANICAL SOLUTION

The three-story, 59,000-square-foot building is of masonry construction with an over-all resistance value of $R = 14$. Its windows are double glazed and the main entrance uses the air-lock design.

The 8,000 square feet of liquid-cooled, flat-plate collectors are arranged in a single plane, 80 feet by 100 feet, and tilted to 35 degrees from the horizontal. The heat removal medium is paraffinic oil. The storage is composed of a 10,000-gallon and a 5,000-gallon tank, each equipped with a heat exchanger. The building uses two separately controlled heating zones, heated by fan-coil units in the forced air system. Cooling is provided by a solar powered absorption chiller. Solar energy provides 95 percent of the space heating and 70 percent of the space cooling needs, with the make-up energy supplied by the central steam plant.

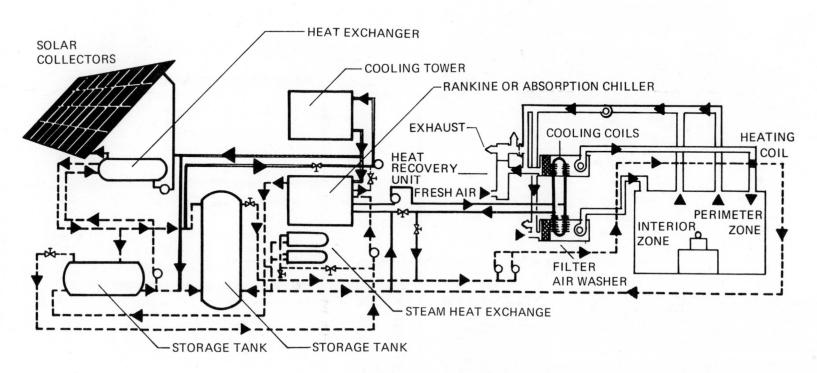

SOLAR SYSTEM SCHEMATIC

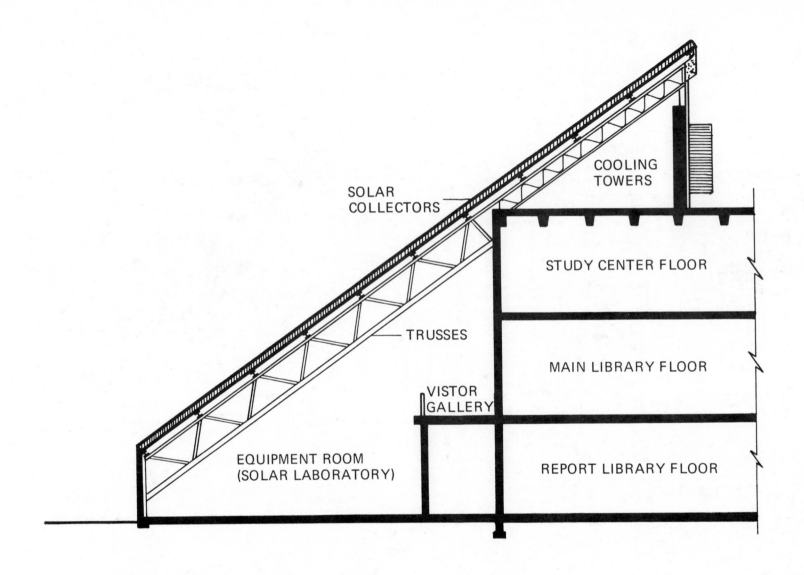

SOLAR COLLECTORS

COOLING TOWERS

TRUSSES

VISTOR GALLERY

STUDY CENTER FLOOR

MAIN LIBRARY FLOOR

REPORT LIBRARY FLOOR

EQUIPMENT ROOM (SOLAR LABORATORY)

CROSS SECTION

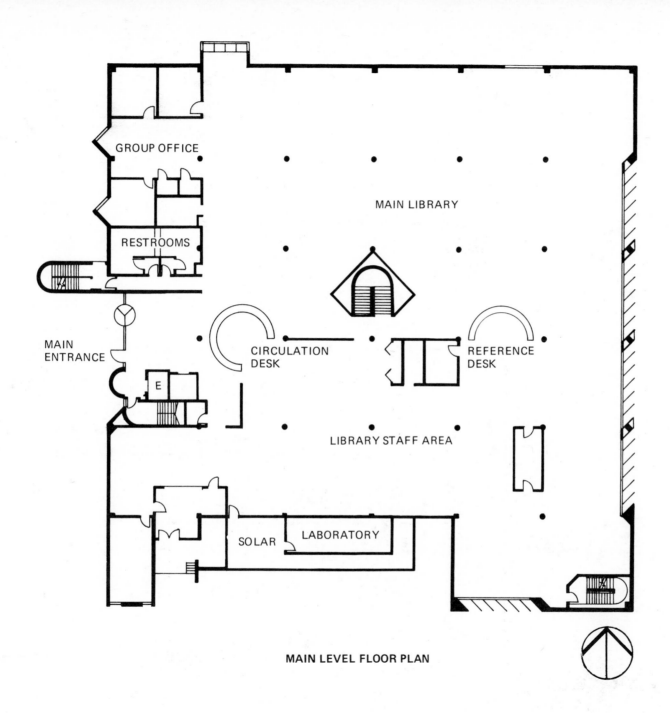

GROUP OFFICE

MAIN LIBRARY

RESTROOMS

MAIN
ENTRANCE

E

CIRCULATION
DESK

REFERENCE
DESK

LIBRARY STAFF AREA

SOLAR   LABORATORY

MAIN LEVEL FLOOR PLAN

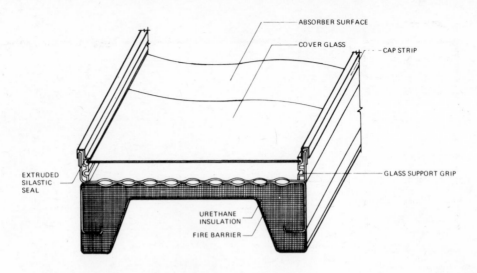

ABSORBER SURFACE

COVER GLASS

CAP STRIP

EXTRUDED
SILASTIC
SEAL

GLASS SUPPORT GRIP

URETHANE
INSULATION

FIRE BARRIER

SOLAR COLLECTOR DETAIL

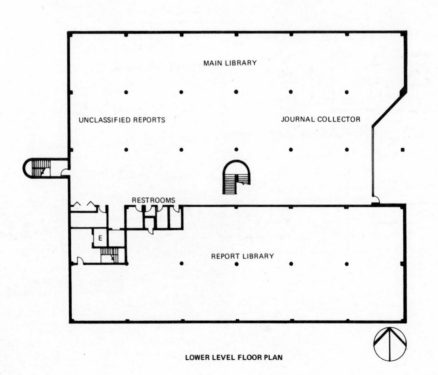

MAIN LIBRARY

UNCLASSIFIED REPORTS

JOURNAL COLLECTOR

RESTROOMS

E

REPORT LIBRARY

LOWER LEVEL FLOOR PLAN

Southwest elevation

Northwest elevation

Single plane collectors

87

TEMPE UNION HIGH SCHOOL

LOCATION: Tempe, Arizona.

LATITUDE: 33.5 degrees N.

REGION: Hot Arid.

OWNER AND BUILDER: Temple Union High School District 213.

MECHANICAL SOLUTION

The school has a solar system which provides the energy required for space and domestic water heating. The rooftop-mounted collectors face south and are tilted to 32 degrees from the horizontal. Heated water from the storage tank is pumped through heat exchangers, thereby transferring the thermal energy to a four-pipe heating/cooling system. Forced air is distributed by the air handlers to the various locations in the building.

Energy conservation features include earth berming at perimeter walls, concentration of building mass, site orientation in accordance with sun angles and the extensive use of insulation.

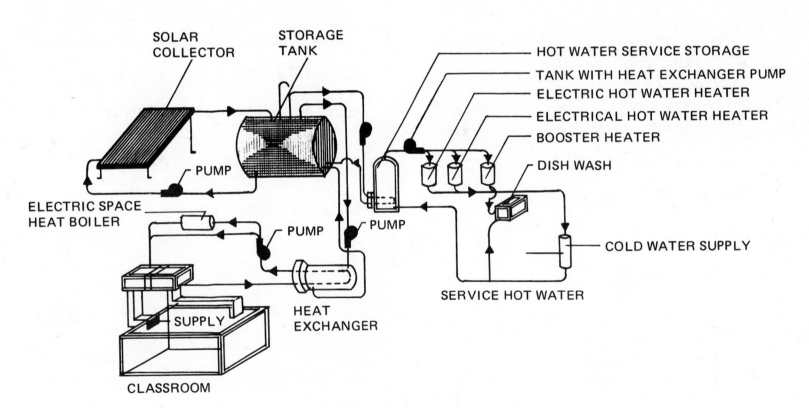

SOLAR COLLECTOR
STORAGE TANK
HOT WATER SERVICE STORAGE
TANK WITH HEAT EXCHANGER PUMP
ELECTRIC HOT WATER HEATER
ELECTRICAL HOT WATER HEATER
BOOSTER HEATER
DISH WASH
PUMP
ELECTRIC SPACE HEAT BOILER
PUMP
PUMP
COLD WATER SUPPLY
SUPPLY
HEAT EXCHANGER
SERVICE HOT WATER
CLASSROOM

SOLAR SYSTEM SCHEMATIC

Section 4. Hot Humid Region

The buildings presented in this section are the following:

George A. Towns Elementary School. Atlanta, Georgia.

The Shenandoah Community Center. Shenandoah, Georgia.

Charlotte Memorial Hospital. Charlotte, North Carolina.

Trinity University. San Antonio, Texas.

GEORGE A. TOWNS ELEMENTARY SCHOOL

LOCATION: Atlanta, Georgia.
LATITUDE: 33.8 degrees N.
REGION: Hot Humid.
ARCHITECT: Burt, Hill & Associates, Butler, Pennsylvania.
MECHANICAL ENGINEER: Dubin, Mindell and Bloome, New
   York, New York.
PRIME CONTRACTOR: Westinghouse Electric Corporation.

MECHANICAL SOLUTION

This one-story, 32,000-square-foot school was retrofitted to use solar energy for space heating and cooling and for domestic hot water heating. To do this, 10,000 square feet of water-cooled, flat-plate collectors were installed on the flat roof and tilted to 45 degrees from the horizontal. The heat removing medium is water with corrosion inhibitors. The collector is automatically drained each evening and filled with nitrogen. To the south of eleven of the twelve collector arrays, there is a reflector sheet

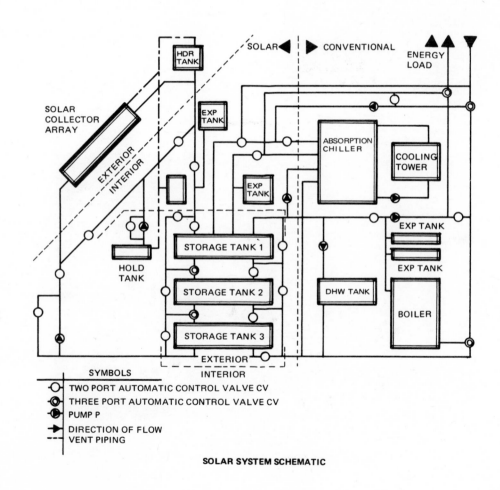

SOLAR SYSTEM SCHEMATIC

Main entrance

which slopes 36 degrees from the horizontal and faces north. This increases the amount of energy collected, while at the same time shading the roof.

The storage is composed of three tanks with a total capacity of 45,900 gallons. In the summer, one of these tanks contains chilled water.

Cooling is provided by a solar assisted, 100-ton absorption chiller. The solar energy system yields 50 percent of the space heating, 60 percent of the space cooling and 80 percent of the domestic hot water. The make-up energy is supplied by an existing gas fired furnace.

Roof mounted collectors Photo by Architect

92

THE SHENANDOAH COMMUNITY CENTER

LOCATION: Shenandoah, Georgia.
LATITUDE: 33.6 degree N.
REGION: Hot Humid.
ARCHITECT: Taylor & Collum, Atlanta, Georgia.
MECHANICAL ENGINEER: Wright Engineering Associates,
 Atlanta, Georgia.

SPACIAL SOLUTION

This multi-purpose community center was designed for the new town of Shenandoah. The 55,000-square-foot, two-story building includes offices, meeting rooms, conventional recreation facilities, a medium-sized skating rink and a heated, outdoor, olympic-sized swimming pool.

The building's shell is of reinforced concrete, with all exterior walls heavily bermed. There are numerous energy saving fea-

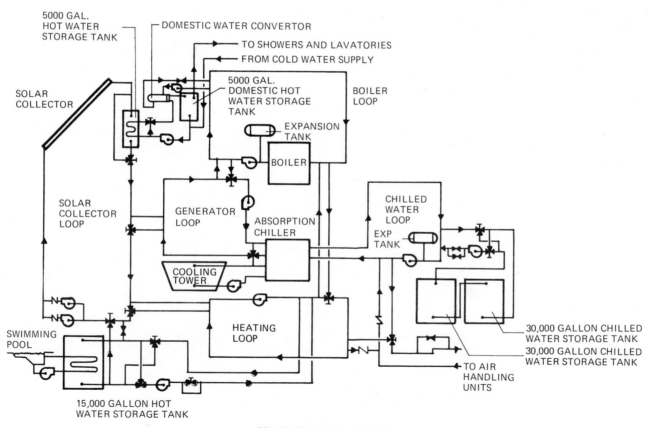

SOLAR SYSTEM SCHEMATIC

tures implemented in the center. All windows are single glazed and screened, and they comprise less than 10 percent of the exterior surface area.

MECHANICAL SOLUTION

The solar energy system was designed to provide space heating and cooling and domestic hot water.

There are 10,500 square feet of water-cooled, flat-plate collectors, lined up in nine rows, mounted on the flat roof and tilted to 45 degrees from the horizontal. The heat removing agent is water, with freezing prevented at night by periodic circulation of hot water from the storage through the collector. The amount of water needed to carry this out is small, and collector heat losses are reduced by the double cover plate. Daytime circulation is maintained by a 20-horsepower centrifugal pump. There is a north-facing, polished aluminum reflector sheet tilted to 36 degrees from the horizontal directly to the south of eight of the nine collector rows. This increases the collector efficiency during the summer. The storage element is composed of five tanks: the 15,000-gallon, 5,000-gallon and 500-gallon tanks are used for storing hot water, and the two 30,000-gallon tanks store chilled water.

The rooms are heated by a fan-coil system and are cooled by a solar assisted 100-ton, lithium bromide absorption chiller.

Solar heated water is used to melt the top ice layer in the skating rink (this is a procedure frequently followed in resurfacing the ice). The olympic-sized swimming pool is heated in the spring and fall.

The community center uses solar energy to provide 95 percent of the space heating, 70 percent of the space cooling and 95 percent of the domestic hot water. The make-up energy is provided by a gas heater.

CHARLOTTE MEMORIAL HOSPITAL

LOCATION: Charlotte, North Carolina.
LATITUDE: 35.2 degrees N.
REGION: Hot Humid.
OWNER AND BUILDER: Charlotte Hospital and Medical Center.

MECHANICAL SOLUTION

The building has 65,000 square feet of floor area distributed among six stories. The envelope has external shading devices and limited glass areas, in an effort to reduce heat gains and losses. In addition to these features, the hospital makes extensive use of insulation. The 4,020 square feet of water-cooled, flat-plate collectors are mounted on the roof area over the health and education facility. The mechanical system was designed to make maximum use of the heat generated by the lights and the people within the structure.

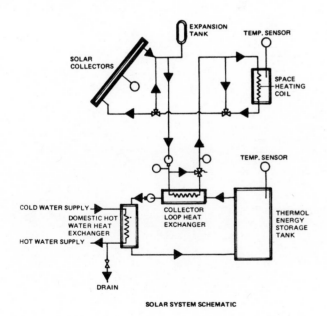

SOLAR SYSTEM SCHEMATIC

TRINITY UNIVERSITY

LOCATION: San Antonio, Texas.
LATITUDE: 29.5 degrees N.
REGION: Hot Humid.
SOLAR SYSTEM DESIGNER AND ENGINEER: Bridges & Paxton, Albuquerque, New Mexico.

MECHANICAL SOLUTION

The University's fuel bill escalated from 1973 to 1975 by 360 percent. This prompted school officials to implement the solar energy system which serves six dormitories and a physical education center, with a total combined floor area of 284,900 square feet.

The 16,000 square feet of concentrated, water-cooled, tracking collectors were mounted on the roof of the gymnasium. The tracking is accomplished by photovoltaic cells; when a shadow

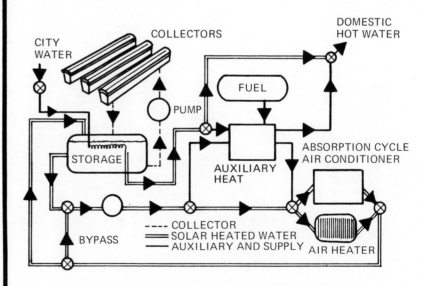

SOLAR SYSTEM SCHEMATIC

95

is cast on the cell, the electrical difference is recorded by an electric circuit which activates the pulley mechanism that moves the collectors. On clear days, the collector generates temperatures of 240 degrees F for a duration of seven to eight hours.

The solar energy system provides 90 percent of the space heating, water heating and space cooling. The make-up energy is provided by a gas fired furnace.

Southwest elevation

Southeast elevation

Northwest elevation

PART II
RESIDENTIAL BUILDINGS

Section 5. Cool Region

The buildings presented in this section are the following:

The Bloomington House. West Bloomington, Minnesota.

Sandy Pines Campground Convenience Center. Hopkins, Michigan.

The Brown House. Ithica, Michigan.

THE BLOOMINGTON HOUSE

LOCATION: West Bloomington, Minnesota.
LATITUDE: 45 degrees N.
REGION: Cool.
ARCHITECT: Marlin Grant, Bloomington, Minnesota.
ENGINEER: Honeywell Inc., Chicago, Illinois.

MECHANICAL SOLUTION

This project involves the application of solar space and domestic water heating to a single family detached unit that has 1,215 square feet of floor area. This wood framed house uses 2 inches of styrofoam sheathing for insulation in the walls, with a total resistance value of $R=25$. The ceiling has 10 inches of Celulose fiberglass insulation and its total resistance value is $R=36$. The 378 square feet of liquid-cooled, flat-plate collectors are mounted on the steep, south-facing portion of the roof. The heat removing agent is circulated through the collectors and to the heat exchanger unit in the 1,000-gallon tank located in the basement. From this tank, heated water is circulated to the fan-coils in the primary supply ducts. The domestic hot water is preheated by the collectors and is stored in a secondary tank that serves as a conventional water heater. The solar system provides 55 percent of the space heating and 68 percent of the domestic hot water. The back-up energy is supplied by a gas fired furnace.

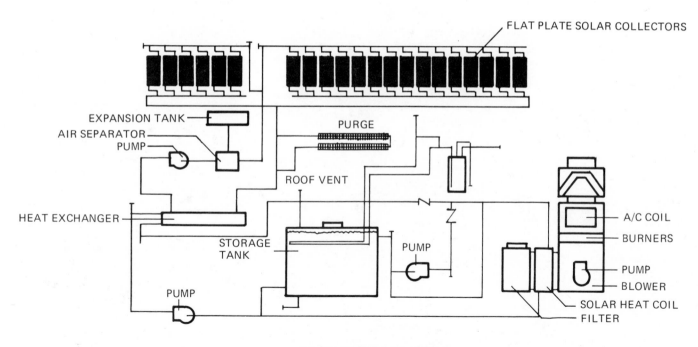

SOLAR SYSTEM SCHEMATIC

Photo by Honeywell, Inc.

102

SANDY PINES CAMPGROUND CONVENIENCE CENTER

LOCATION: Hopkins, Michigan.
LATITUDE: 44 degrees N.
REGION: Cool.
ARCHITECT: Rodney Wright, Chicago, Illinois.

SPACIAL SOLUTION

This one-story, 2,500-square-foot convenience center is part of the 812-acre private camping resort known as Sandy Pines. The building is all masonry and uses 2 inches of styrofoam insulation in the walls and 9 inches of fiberglass insulation in the ceiling. In addition to this, the east, west and south walls are heavily bermed. The windows are double glazed and are used in the north wall as a clerestory. The entrances to the convenience center are recessed to provide protection from the winds and rains.

Southeast elevation

MECHANICAL SOLUTION

The solar system was designed to provide space heating and domestic hot water. There are 800 square feet of air-cooled, flat-plate collectors mounted in two rows; one on the building's south end and one on the north end. Both rows are tilted to 53 degrees from the horizontal. The collectors form an integral part of the exterior architecture of the building. A 1-horsepower blower circulates the heated air from the collectors to the storage tank. This tank, made of concrete block, contains 100 tons of 2-inch diameter stones and 30 pieces of 1/2-inch diameter copper tubing. The tubes, which are on top of the rocks, serve to preheat the domestic hot water used in the center. A 3/4-horsepower blower circulates the heated air from the storage tanks to the rooms. The solar system provides 65 percent of the space heating needs, and the make-up energy is supplied by electric coils in the duct.

East elevation

103

Thirty pieces of 1/2 inch copper tubing

Storage 100 tons of stone

104

THE BROWN HOUSE

LOCATION: Ithica, Michigan.
LATITUDE: 44.5 Degrees N.
REGION: Cool.
ARCHITECT: Larry Brown, Ithica, Michigan.

MECHANICAL SOLUTION

This three-bedroom wood framed house uses solar energy to provide the space heating needs. However, the most innovative and attractive feature of the house is the use of 4,000 beer cans cut in half and attached in a honeycomb arrangement to form the collector's absorbing plate. The cans are covered with a single glazed cover and they face directly south. A blower circulates the heated air from the collectors to the two storage tanks located in the basement. Each tank contains 25 tons of stones. The warm air is allowed to rise from the tanks through the duct system and into the rooms. This flow is controlled by thermostatically operated louvers. The back-up energy is provided by electric heaters.

South elevation

Northwest elevation

Section 6. Temperate Region

The buildings presented in this section are the following:

The Skytherm House. Atascadero, California.

The Kittle House. Gaviota, California.

The Ward House. Davis, California.

The Sundown House. Sea Ranch, California.

The Wood House. Colorado Springs, Colorado.

The Phoenix Colorado Springs House. Colorado Springs, Colorado.

19th Street Solar Housing. Boulder, Colorado.

The Lawrence Fellows Solar Fence. Bloomfield, Connecticut.

Pyramidal Optics Solar House. Stamford, Connecticut.

The Barber House. Guilford, Connecticut.

The Jantzen House I. Carlyle, Illinois.

The Jantzen House II. Carlyle, Illinois.

The Thomason Solar House No. 6. Malboro, Maryland.

The Zwillinger House. New Boston, New Hampshire.

The Kelbaugh House. Princeton, New Jersey.

The Evans House. East Hampton, New York.

The Wenning House. La Grangeville, New York.

The Barbash House. Quoque, New York.

The Engle House. Wagoner, Oklahoma.

The Bishoprick House. Salem, Oregon.

The People Space House. Windham, Vermont.

THE SKYTHERM HOUSE

LOCATION: Atascadero, California.
LATITUDE: 35 degrees N.
REGION: Temperate.
ENGINEER, DESIGNER AND OWNER: H. R. Hay, Skytherm
    Process & Engineering, Los Angeles, California.
ARCHITECT: Kenneth Haggard, California Polytech.

SPACIAL SOLUTION

This one-story, seven-room, 1,140-square-foot house has the appearance of any conventionally heated house— but it is conventional *only* in appearance. The roof of the house contains four water-filled bags, 58 feet long, 8 feet wide and 1 foot deep, with a combined capacity of 6,300 gallons. There is a single glazed cover plate above the bags, which rest on a black plastic sheet that covers the steel deck used as the ceiling. During the winter

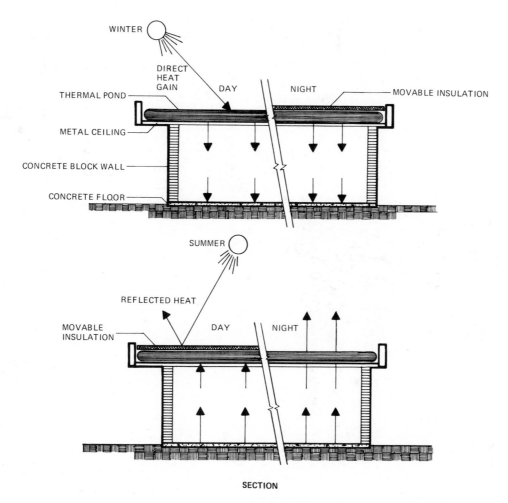

SECTION

nights, the bags are covered with nine horizontally sliding insulated panels. In the morning, the panels automatically slide into an open position, leaving the bags exposed to absorb the solar radiation. At night, the collected thermal energy is released to the interior spaces. When the temperature of the water in the bags reaches 85 degrees, they are able to heat the house through four sunless days. During the summer days, the bags are covered to prevent direct solar heat gain. They absorb heat from the rooms, and at night the panels are opened to allow the release of the absorbed heat.

Skytherm provides 100 percent of the space heating and cooling needs. No back-up energy source is used, although an electric heater is available. A 1/3-horsepower motor is used a total of five minutes a day to slide the panels, but this can also be done manually. Five similar units have been built in Selma, California.

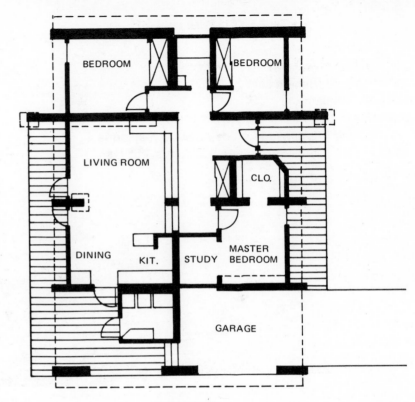

FLOOR PLAN

Front elevation

THE KITTLE HOUSE

LOCATION: Gaviota, California.
LATITUDE: 34.5 degrees N.
REGION: Temperate.
ARCHITECT AND SOLAR ENGINEER: Zomeworks,
    Albuquerque, New Mexico.

SPACIAL SOLUTION

This 2,500-square-foot house is located on a working ranch in Las Cruses, north of Santa Barbara. The house and the solar system were designed to provide space and domestic water heating. This combination is virtually independent of fossil fuels as a back-up energy source. The cluster of ten zomes is sheathed with redwood that was recycled from old wine vats. The exterior walls and roofs are insulated with 4 inches of fiberglass. The interior walls and the floor slab are of massive construction, which enables them to store the solar radiation that reaches the interior spaces. There are 300 square feet of skylids to control the solar penetration. (A complete definition of skylids, with drawings, can be found in Section II in the discussion of the airport terminal in Aspen, Colorado.

The thermal qualities of the house provide 75 percent of the space heating needs. The remaining heating needs are supplied by four wood burning fire stoves. The large thermal capacity of the walls and floor eliminates the need for space cooling, since large thermal gains do not take place during the summer months.

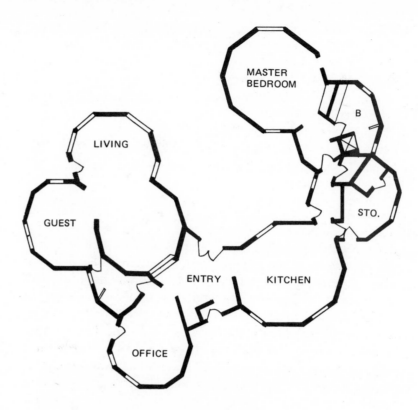

FLOOR PLAN

THE WARD HOUSE

LOCATION: Davis, California.
LATITUDE: 38.5 degrees N.
REGION: Temperate.
ARCHITECT: Michael N. Cobertt, Davis, California.
SOLAR ENGINEER: Natural Heating Systems, Davis, California.

SPACIAL SOLUTION

This one-story, 1,164-square-foot house has three bedrooms and is located in a residential development. It was erected on an insulated concrete slab that acts as a heat sink. The walls and ceiling are insulated to $R=11$ and $R=19$, respectively. There are no windows on the east or west walls. The north wall has four small windows and the south wall has a total of 175 square feet of glass area. The windows on the wall allow direct heat gains during the winter. During the summer, they are shaded by the roof overhang. At night, the windows are covered with a 1-inch thick styrofoam panel to prevent heat losses. The main entrance is recessed and it is from the west. A secondary entrance is located in the east wall.

MECHANICAL SOLUTION

The house has 204 square feet of flat-plate, water-cooled collectors, integrated with the tile roof and tilted to 60 degrees from the horizontal. An overhang placed over the collectors effectively shades them during the summer, thereby preventing the heat build-up from damaging the collectors. Water is circulated through the collectors and to the storage tank by "thermosyphoning," without the aid of pumps. The attic contains the 209-gallon steel storage tank. Heat is supplied to a room by six separately controlled radiant floor panels. A 1/2-horsepower pump draws water from the top of the storage tank and circulates it through the 780 feet of copper tubes imbedded in the floor slab. The pump is activated by a thermostat at the top of the tank; the slab absorbs the heat and releases it to the living spaces. Domestic water is preheated by a heat exchanger in the storage tank.

The solar system provides over 65 percent of the space and domestic water heating needs, with the make-up heat being provided by a Franklin stove and a gas fired wall furnace.

Due to the inherent thermal qualities of the construction, natural nocturnal cooling is sufficient to maintain a comfortable interior temperature during the summer.

Integrated collectors

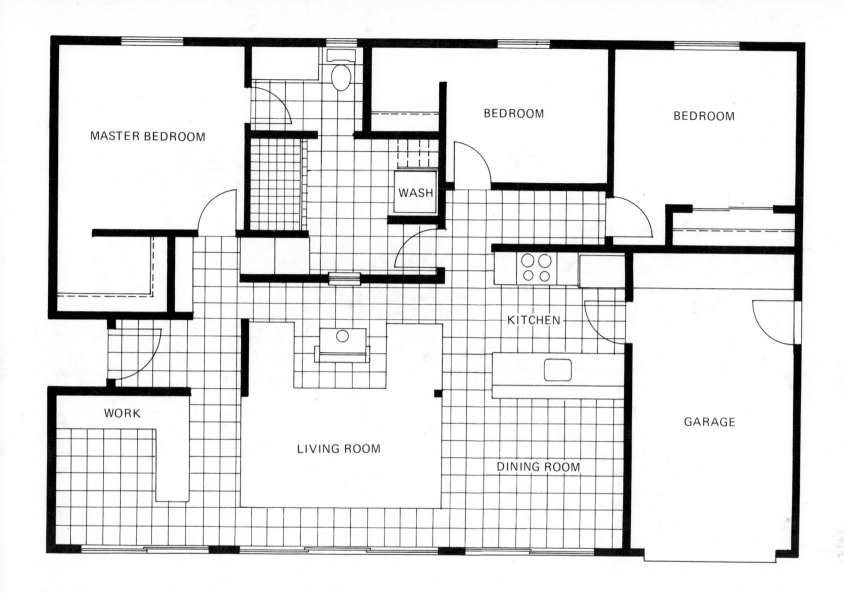

MASTER BEDROOM

BEDROOM

BEDROOM

WASH

WORK

LIVING ROOM

KITCHEN

DINING ROOM

GARAGE

FLOOR PLAN

0   1.5   3.0   6 FEET

113

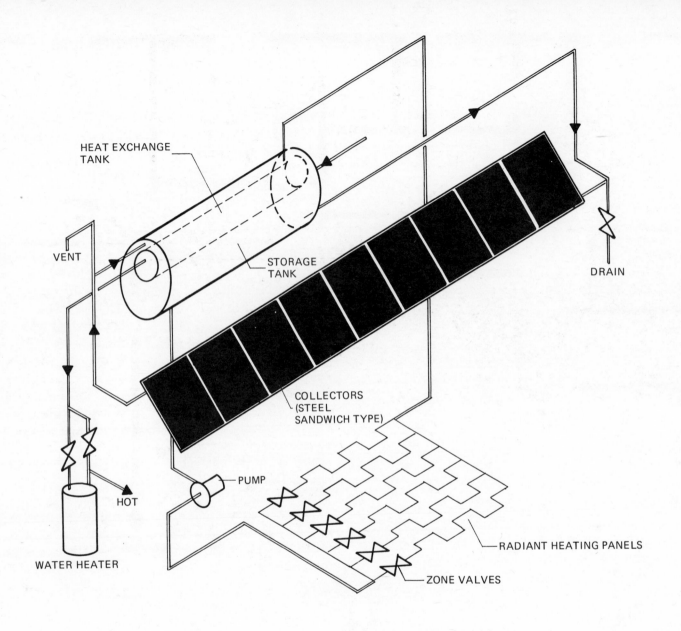

HEAT EXCHANGE TANK

VENT

STORAGE TANK

DRAIN

COLLECTORS (STEEL SANDWICH TYPE)

HOT

PUMP

RADIANT HEATING PANELS

WATER HEATER

ZONE VALVES

SOLAR SYSTEM SCHEMATIC

THE SUNDOWN HOUSE

LOCATION: Sea Ranch, California.
LATITUDE: 39 degrees N.
REGION: Temperate.
ARCHITECT AND SOLAR DESIGNER: David Wright, Sea
    Ranch, California.
SOLAR CONSULTANTS: M. Chalom and K. Haggard, San
    Francisco, California.

SPACIAL SOLUTION

This house has 1,146 square feet of floor area distributed among one floor and a loft tower, plus a detached garage and studio. The wood framed house is built of stone masonry and is shaped around a sunken and open patio. The northwest, northeast and part of the southeast walls are totally covered with earth berms which flow onto the sod roof. This provides extremely high insulation as well as protection from the prevailing winds. The southeast wall is composed of 376 square feet of double glazed windows, tilted to 75 degrees from the horizontal. This configuration allows the winter sun to penetrate the interior spaces where the walls and floor store the thermal energy. Sixty-four square feet of water-cooled, flat-plate collectors have been integrated with the southeast wall to provide the domestic hot water. The water flows directly from the collector into an 80-gallon storage tank.

The energy conscious design concepts have been materialized and detailed to such a refined state that the house only makes occasional use of the wood stove or the fireplace to provide additional heating.

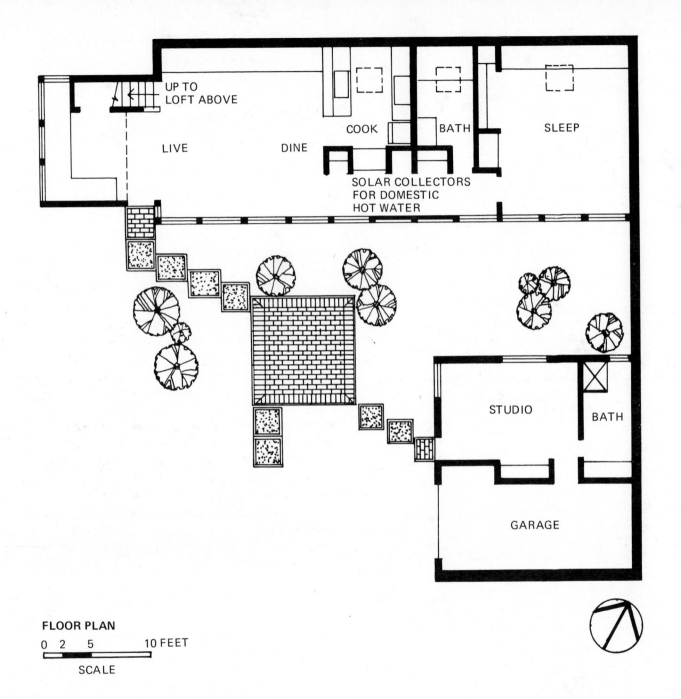

UP TO
LOFT ABOVE

LIVE

DINE

COOK

BATH

SLEEP

SOLAR COLLECTORS
FOR DOMESTIC
HOT WATER

STUDIO

BATH

GARAGE

FLOOR PLAN

0 2 5 10 FEET

SCALE

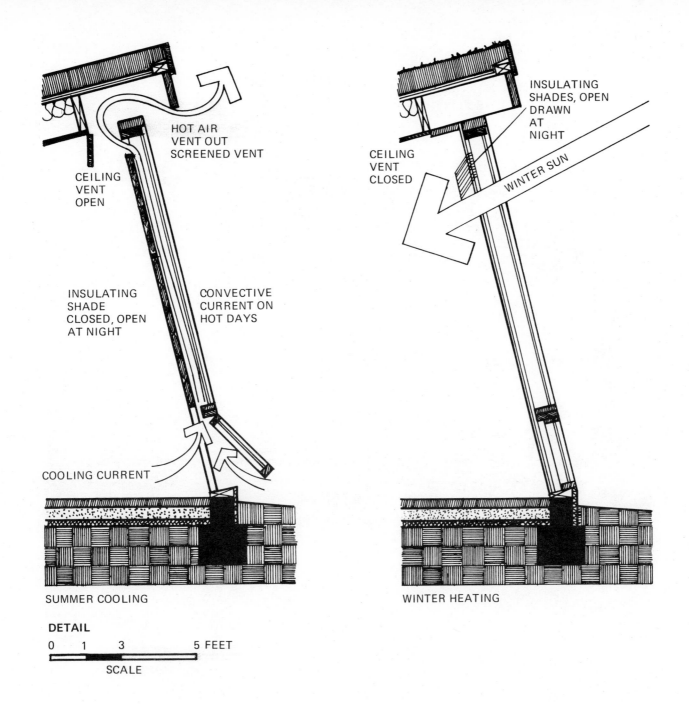

HOT AIR
VENT OUT
SCREENED VENT

CEILING
VENT
OPEN

INSULATING
SHADE
CLOSED, OPEN
AT NIGHT

CONVECTIVE
CURRENT ON
HOT DAYS

COOLING CURRENT

SUMMER COOLING

DETAIL

0 1 3 5 FEET

SCALE

INSULATING
SHADES, OPEN
DRAWN
AT
NIGHT

CEILING
VENT
CLOSED

WINTER SUN

WINTER HEATING

117

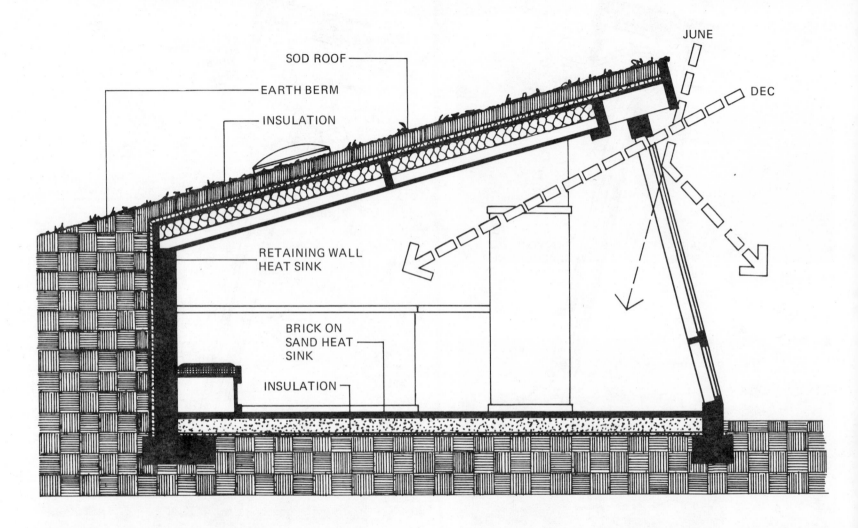

SOD ROOF

EARTH BERM

INSULATION

RETAINING WALL
HEAT SINK

BRICK ON
SAND HEAT
SINK

INSULATION

JUNE

DEC

SECTION

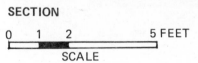

0   1   2   5 FEET
SCALE

Southeast elevation

THE WOOD HOUSE

LOCATION: Colorado Springs, Colorado.
LATITUDE: 39 degrees N.
REGION: Temperate.
ARCHITECT AND SOLAR ENGINEER: P. O. Wood, Colorado
    Springs, Colorado.

SPACIAL SOLUTION

This two-story house has a total floor area of 2,400 square feet
and a basement. It has three bedrooms, three baths, cathedral
ceilings, a sun deck and a two-car garage. The walls are insulated with 3-1/2 inches of fiberglass, and the ceiling with 6 inches of blown-in insulation. The house is oblong in plan, with
the longer side facing 10 degrees east of south.

South elevation

Interior

119

MECHANICAL SOLUTION

The 750 square feet of flat-plate, water-cooled collectors are mounted as an integral part of the south-facing roof and constitute the main visual element of the building. The collectors, which are manufactured by Solaris, are tilted to 45 degrees from the horizontal. They use trickling water, pumped up to the collectors by a 1/3-horsepower pump, to remove the collected energy. The water contains no antifreeze or corrosion inhibitors and it is drained into the storage tank at the end of the day. The 1,800-gallon storage tank is surrounded by 40 tons of stones. A 1/2-horsepower blower circulates the heated air from the storage bin into the rooms. The domestic hot water is preheated by a heat exchanger in the water storage tank.

During the summer nights, water is cooled by circulating it down the north side of the roof. Once cooled, it is returned to the storage tank, where it absorbs heat from the stones. Air is circulated from the rooms through the storage bin, effectively cooling it. The air is then recirculated to the rooms, where it absorbs heat.

The system provides 75 percent of the space heating and domestic hot water needs. The make-up energy is supplied by a gas fired furnace.

THE PHOENIX COLORADO SPRINGS HOUSE

LOCATION: Colorado Springs, Colorado.
LATITUDE: 39 degrees N.
REGION: Temperate.
ARCHITECT: Design Group Architects, Colorado Springs, Colorado.
SOLAR ENGINEER: D. M. Jardine, Colorado Springs, Colorado.

SPACIAL SOLUTION

This two-story, 2,300-square-foot house was built as the result of the joint effort of the Colorado Springs City government, the Colorado building industry and the National Science Foundation. The exterior architecture is dominated by the two collector arrays that shade the south-facing windows. Wind protection fins were built around the house.

The ceiling is insulated with 6 inches of mineral wool, and the walls with 3-1/2 inches of fiberglass insulation.

MECHANICAL SOLUTION

The solar system provides space and water heating and is equipped with 810 square feet of liquid-cooled, flat-plate collectors, tilted to 55 degrees from the horizontal. The roof has white quartz to reflect additional radiation into the collectors. The heat removal agent is a combination of antifreeze and water, flowing from the collector to the heat exchanger in the 7,000-gallon storage tank. Heat is distributed to the room by fan-coil units in the forced air system. A heat pump is used when the water temperature in the tank drops below 90 degrees F. The domestic hot water is preheated in the tank.

Summer cooling is provided by the heat pump and natural ventilation.

North elevation

Southwest elevation

19TH STREET SOLAR HOUSING

LOCATION: Boulder, Colorado.
LATITUDE: 40 degrees N.
REGION: Temperate.
ARCHITECT: Joint Venture, Inc., Boulder, Colorado.
SOLAR ENGINEER: Dr. Jan. F. Kreider, Boulder, Colorado.

SPACIAL SOLUTION

This project consists of eight units with a total of 8,475 square feet of floor area, distributed among four floors. The units vary from one another in plan and section. The upper two units have two stories and a sundeck, while the lower units are split level. All the units have four bedrooms, 1-1/2 bathrooms, a kitchen, a dining area and a living room.

The construction is wood frame and the walls are insulated to $R=20$ and the roof to $R=40$. The entrances were designed with air-lock vestibules to reduce heat losses. All the windows are double glazed.

MECHANICAL SOLUTION

The solar system was designed to provide space and domestic water heating. On the south facade, 860 square feet of compound parabolic, concentrating, liquid-cooled collectors are integrated into the roof structure. The heat removal agent is circulated through the collectors and the heat exchangers in the 1,700-gallon storage tank by a centrifugal pump. Heat is distributed to the apartments through baseboard convectors. The domestic hot water is preheated in the storage tank. Cooling is achieved through natural ventilation.

The solar system provides 74 percent of the space and water heating needs, and the make-up energy is supplied by a gas fired water heater.

121

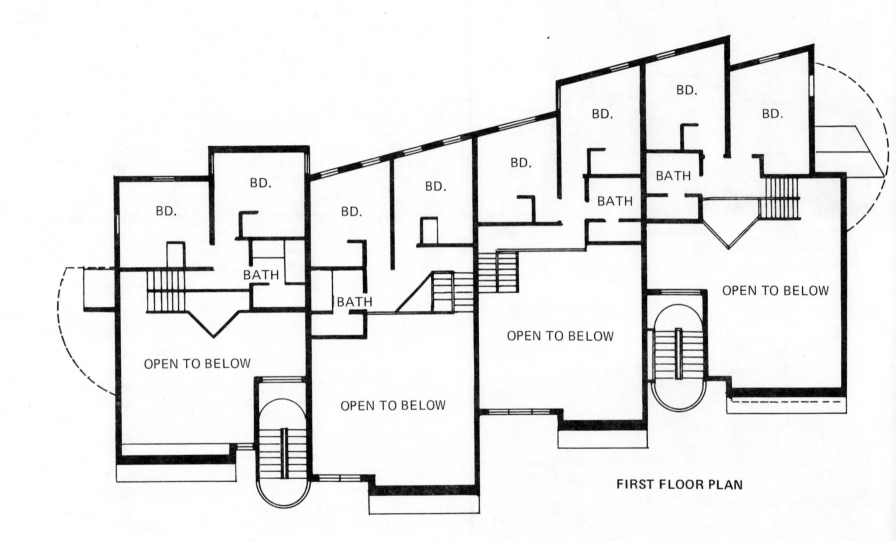

BD.

BD.

BD.

BD.

BD.

BD.

BD.

BD.

BATH

BATH

BATH

BATH

BATH

OPEN TO BELOW

OPEN TO BELOW

OPEN TO BELOW

OPEN TO BELOW

OPEN TO BELOW

FIRST FLOOR PLAN

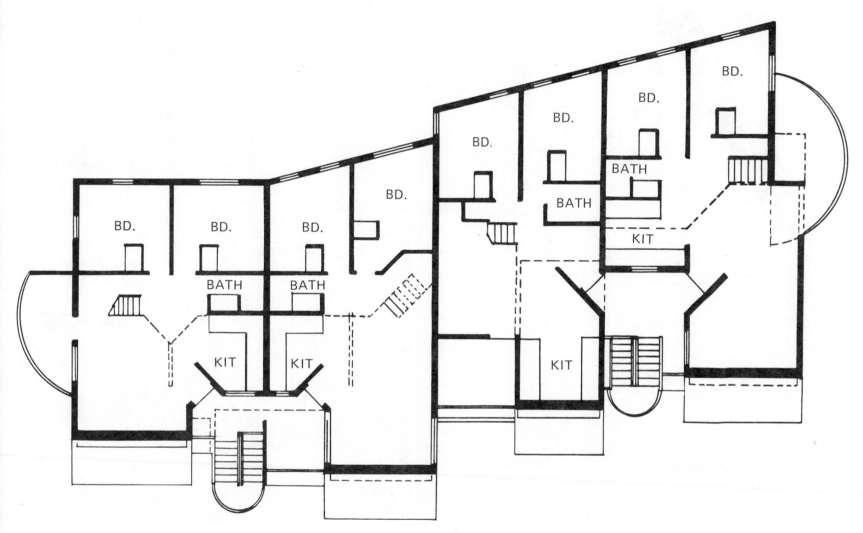

SECOND FLOOR PLAN

123

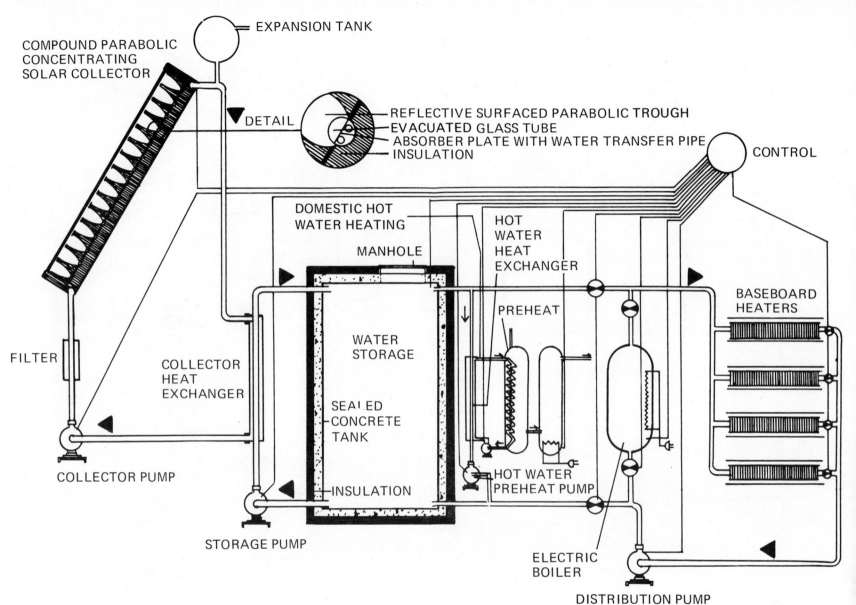

EXPANSION TANK

COMPOUND PARABOLIC
CONCENTRATING
SOLAR COLLECTOR

DETAIL

REFLECTIVE SURFACED PARABOLIC TROUGH
EVACUATED GLASS TUBE
ABSORBER PLATE WITH WATER TRANSFER PIPE
INSULATION

CONTROL

DOMESTIC HOT
WATER HEATING

HOT
WATER
HEAT
EXCHANGER

MANHOLE

BASEBOARD
HEATERS

PREHEAT

FILTER

WATER
STORAGE

COLLECTOR
HEAT
EXCHANGER

SEALED
CONCRETE
TANK

COLLECTOR PUMP

INSULATION

HOT WATER
PREHEAT PUMP

STORAGE PUMP

ELECTRIC
BOILER

DISTRIBUTION PUMP

SOLAR SYSTEM SCHEMATIC

124

South elevation

Apartment interior

North elevation **Photos by Group Communications**

THE LAWRENCE FELLOWS SOLAR FENCE

LOCATION: Bloomfield, Connecticut.
LATITUDE: 42 Degrees N.
REGION: Temperate.
DESIGNER AND INVENTOR: Lawrence Fellows, Bloomfield, Connecticut.
ENGINEERS: Sidney Berson and Paul Hufford, Bloomfield, Connecticut.

SOLUTION

Although this particular implementation of solar energy is limited to the heating of a swimming pool, it has been included in the book because of its innovative and imaginative characteristics. In his back yard, Mr. Fellows has a 35,000-gallon swimming pool. With the help of two engineer friends, he calculated that the pool needed 100 square feet of collector area. In order to minimize the visual impact that a single collecting plane of this dimension would have on the neighborhood, four 4-foot by 8-foot panels were mounted on the fence. The collector boxes were framed with 1-inch by 6-inch cedar boards. The back of each box was covered with cedar pickets to maintain the fence appearance. However, the pickets left space through which heat would be lost; therefore, the backs of the boxes were lined with polyurethane foam. The cover plate consists of a single sheet of clear polycarbonate plastic.

When not in use, the panels are locked into a vertical position. When in use, the panels pivot on pieces of galvanized pipe to the appropriate angle. Water circulates through finned copper pipes in the inside of the boxes and to the filter unit, which powers the circulation in the solar system.

PYRAMIDAL OPTICS SOLAR HOUSE

LOCATION: Stamford, Connecticut.
LATITUDE: 41 degrees N.
REGION: Temperate.
SOLAR ENGINEERS: G. Fabel and E. M. Wormser, Stamford, Connecticut.

MECHANICAL SOLUTION

This is a retrofit project. The collectors and storage areas were added on to heat the domestic hot water.

The collection element consists of a rectangular funnel, with the top plane tilted to 27 degrees from the horizontal and the side planes slanted 30 degrees inward. This configuration forms a pyramidal structure—hence the name. This pyramidal collector is located on the upper and southern part of the house and forms part of the structure and space defining elements of the house. On the south end, the pyramid is covered with a flat plate, which, like the rest of the interior planes, is covered with reflective mylard. This flap faces north and its angle of inclination can be adjusted to maximize the amount of solar radiation reflected into the collector. At night and when heat collection is not required, the flap is raised to a vertical position. This effectively closes the pyramid's south end and protects the unit from rain, snow, birds, winds and other elements that might cause damage. At the opposite end of the flap, placed either vertically or horizontally, we find the absorber plate. Because of the characteristics of the pyramidal optics collector, the size and weight of the absorber plate is reduced by a factor of 2 to 6, depending on the climate and functional application. Due to the large amount of reflected solar radiation, the collector yields water with temperatures higher than those yielded by the conventional flat-plate collector. This characteristic reduces the volume requirements of the storage tank. The system was installed in 1974 and over the following four years did not encounter any major difficulties.

128

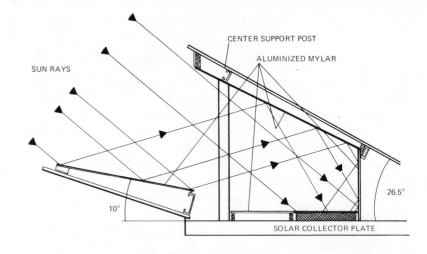

SUN RAYS

CENTER SUPPORT POST

ALUMINIZED MYLAR

26.5°

10°

SOLAR COLLECTOR PLATE

SECTION

Southeast elevation

THE BARBER HOUSE

LOCATION: Guilford, Connecticut.
LATITUDE: 41 degrees N.
REGION: Temperate.
ARCHITECT: Charles W. Moore Associates, Essex,
   Connecticut.
SYSTEM DESIGN: Sunworks, Inc., Guilford, Connecticut.

SPACIAL SOLUTION

This two-story, 1,300-square-foot house is located on a densely wooded site. The spacial and functional organizations were developed based on their energy consumption and harnessing characteristics.

The exterior walls are concrete block, insulated on the outside with 3 inches of sprayed polyurethane foam, which yields a resistance value of $R=20$. The roof is insulated with 6 inches of fiberglass and it yields a resistance value of $R=21$. Inside the house, insulating shutters are used instead of drapes. These are made with 3 inches of rigid insulation and are covered with fabric. The few windows in the house are double glazed. Most of the windows are located on the southern elevation to allow for maximum light and solar heat gains. The windows are shaded during the summer months. A large stone fireplace is the central element of the house and occasionally provides auxiliary heat. A filtered gray-water system is used to flush the toilets.

MECHANICAL SOLUTION

The solar system provides space and domestic water heating. Space cooling is effectively achieved through spacial configuration and natural ventilation.

The southwest side of the house has 400 square feet of liquid-cooled collectors, resting on the roof at a tilt of 57 degrees from the horizontal. The heat removal agent is a combination of water and ethylene glycol. The mixture is circulated through the collectors and the heat exchanger by a 1/12-horsepower pump. There is a cylindrical 2,500-gallon steel storage tank, standing on end behind the fireplace. The tank stores the heat collected from the solar panels and from the water circulated to the fireplace when it is operational. Heat is distributed to the room by a fan-coil system. In addition, there is a 2-foot-deep layer of stones below the floor slab, in which heat extracted from the hot air near the ceiling is stored. The thermal mass inherent in the heavy masonry construction also stores some heat.

The shape of the interior spaces acts to draw air through the house and through the vent at the belvedere. The exhausted air is replaced with cooler air entering at the lower end of the house. This circulation provides cooling during the summer months.

The combination of the solar system and the spacial characteristics provides 70 percent of the space and water heating needs and 100 percent of the space cooling. The make-up energy is supplied by a gas fired heater.

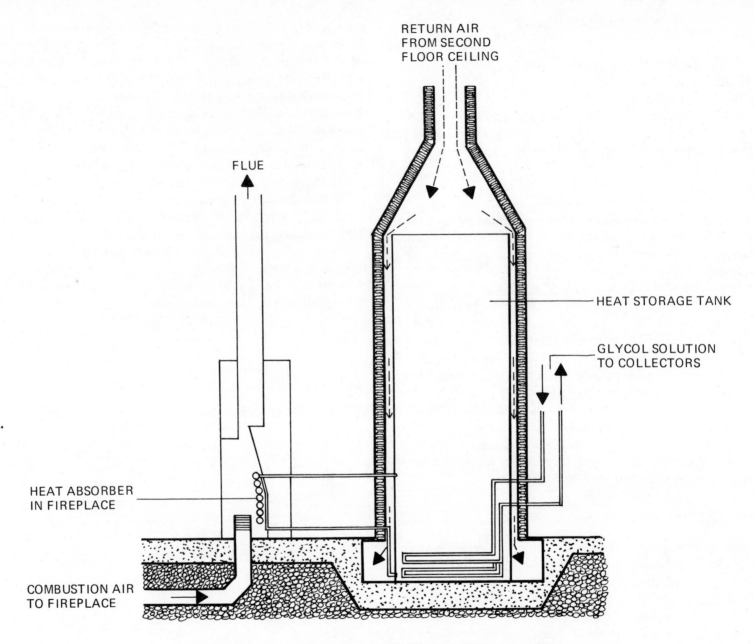

RETURN AIR
FROM SECOND
FLOOR CEILING

FLUE

HEAT STORAGE TANK

GLYCOL SOLUTION
TO COLLECTORS

HEAT ABSORBER
IN FIREPLACE

COMBUSTION AIR
TO FIREPLACE

SECTION THROUGH HEAT STORAGE TANK

130

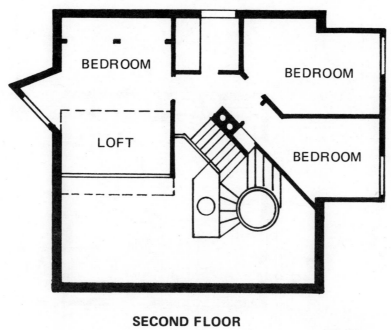

SECOND FLOOR

BEDROOM

BEDROOM

LOFT

BEDROOM

0　　5　　10　　　　20 FEET

SCALE

131

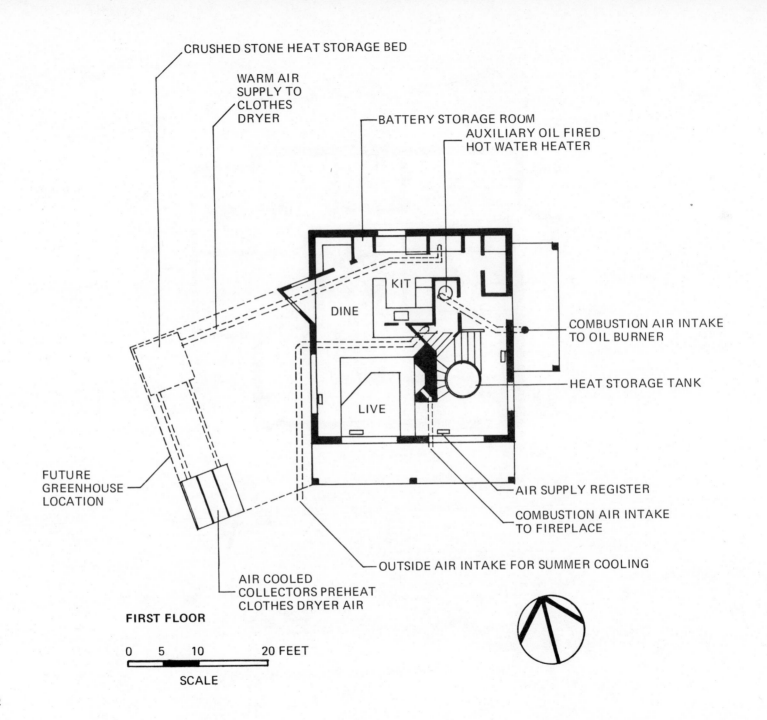

CRUSHED STONE HEAT STORAGE BED

WARM AIR SUPPLY TO CLOTHES DRYER

BATTERY STORAGE ROOM

AUXILIARY OIL FIRED HOT WATER HEATER

KIT

DINE

COMBUSTION AIR INTAKE TO OIL BURNER

HEAT STORAGE TANK

LIVE

FUTURE GREENHOUSE LOCATION

AIR SUPPLY REGISTER

COMBUSTION AIR INTAKE TO FIREPLACE

OUTSIDE AIR INTAKE FOR SUMMER COOLING

AIR COOLED COLLECTORS PREHEAT CLOTHES DRYER AIR

FIRST FLOOR

0   5   10   20 FEET

SCALE

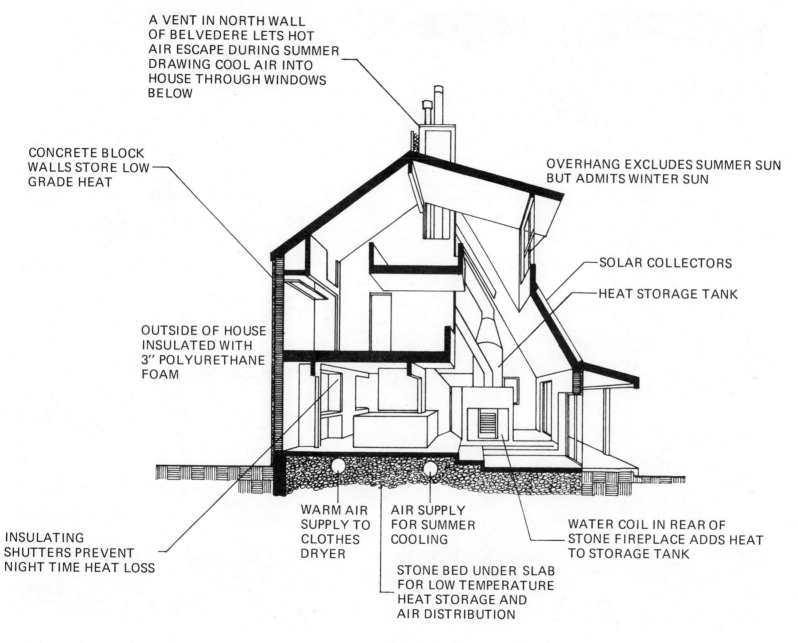

A VENT IN NORTH WALL
OF BELVEDERE LETS HOT
AIR ESCAPE DURING SUMMER
DRAWING COOL AIR INTO
HOUSE THROUGH WINDOWS
BELOW

CONCRETE BLOCK
WALLS STORE LOW
GRADE HEAT

OVERHANG EXCLUDES SUMMER SUN
BUT ADMITS WINTER SUN

SOLAR COLLECTORS

HEAT STORAGE TANK

OUTSIDE OF HOUSE
INSULATED WITH
3'' POLYURETHANE
FOAM

INSULATING
SHUTTERS PREVENT
NIGHT TIME HEAT LOSS

WARM AIR
SUPPLY TO
CLOTHES
DRYER

AIR SUPPLY
FOR SUMMER
COOLING

WATER COIL IN REAR OF
STONE FIREPLACE ADDS HEAT
TO STORAGE TANK

STONE BED UNDER SLAB
FOR LOW TEMPERATURE
HEAT STORAGE AND
AIR DISTRIBUTION

SECTION LOOKING EAST

133

Southeast elevation

Northeast elevation

Dining room and kitchen

Fireplace and storage tank

Stairs to lofts and restrooms

Photos by Robert Perron

136

THE JANTZEN HOUSE I

LOCATION: Carlyle, Illinois.
LATITUDE: 38.5 degrees N.
REGION: Temperate.
DESIGNER, SOLAR ENGINEER AND BUILDER: Michael
 Jantzen, Carlyle, Illinois.

SPACIAL SOLUTION

The main function of this house is to serve as a summer week-end vacation home. Therefore, the house does not require a great deal of heating.

This two-story, two-room, one-bath house has 650 square feet of interior area and a wooden sun deck. The form (exterior and interior) follows the shape of a half-dome, with the curved surface facing north. The structure is of wood laminated arches and it is covered with corrugated steel siding. The second floor has 2 inches of sprayed polyurethane insulation on the walls and ceiling, on top of which a thick layer of fire resistant paint was applied. The first floor has 3 inches of fiberglass insulation covered with wood panelling. Ninety-five percent of the glass area in the house faces south and can be covered with insulating shutters to prevent heat losses.

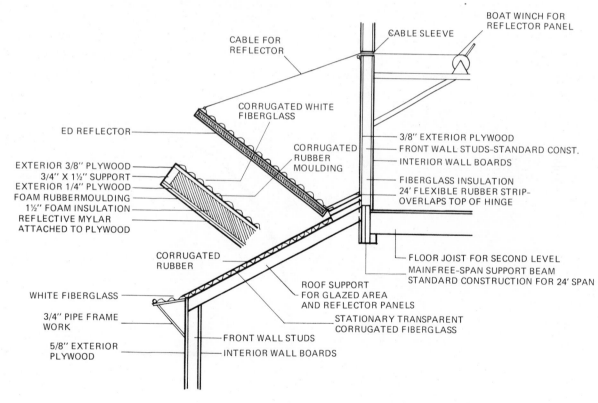

CROSS SECTION OF COLLECTOR AND REFLECTOR

137

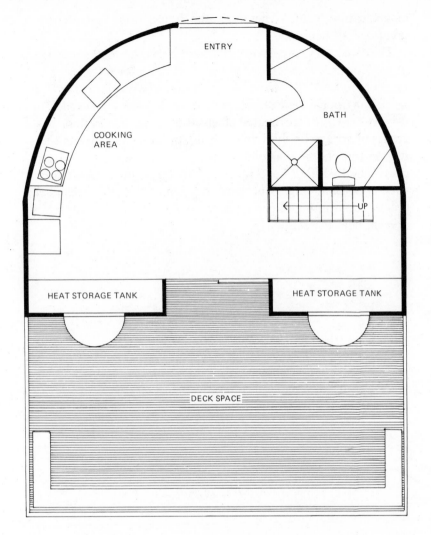

ENTRY

BATH

COOKING
AREA

← UP

HEAT STORAGE TANK | HEAT STORAGE TANK

DECK SPACE

FIRST FLOOR PLAN

0   3'-6''   7'-0''   10'-6''

SCALE

MECHANICAL SOLUTION

The solar system implemented in this house has a unique characteristic in that the collection of the solar energy is done directly by the two storage tanks. On the south end of the roof, there is a transparent panel, tilted to 30 degrees from the horizontal and covered by a movable, insulated, reflective flap. During the day, the flap is tilted up and reflects sun light directly to the storage tanks located below. The angle of reflection can be adjusted manually as needed. The storage tank has a total capacity of 200 gallons and contains a mixture of water and ethylene glycol plus corrosion inhibitors. The tanks are covered, when not in use, by an insulating cushion. The system provides over 50 percent of the specific seasonal heating needs, and a small portable electric heater is used from time to time.

Due to the building's configuration and the efficient design and construction of shading devices, there is no need for artificial cooling. Natural ventilation provides 100 percent of the space cooling needs.

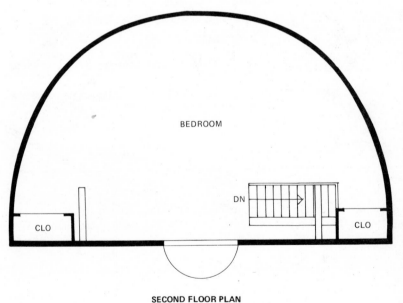

BEDROOM

DN

CLO   CLO

SECOND FLOOR PLAN

138

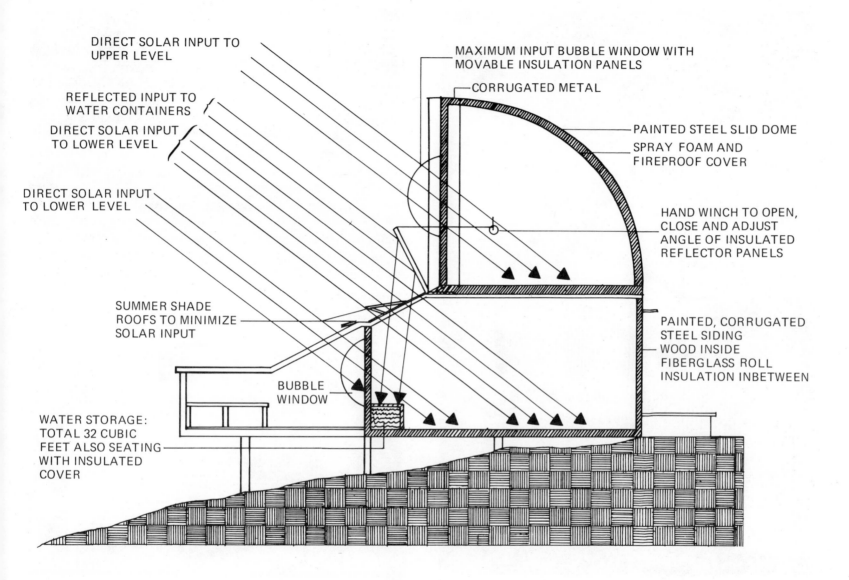

DIRECT SOLAR INPUT TO
UPPER LEVEL

REFLECTED INPUT TO
WATER CONTAINERS

DIRECT SOLAR INPUT
TO LOWER LEVEL

DIRECT SOLAR INPUT
TO LOWER LEVEL

SUMMER SHADE
ROOFS TO MINIMIZE
SOLAR INPUT

WATER STORAGE:
TOTAL 32 CUBIC
FEET ALSO SEATING
WITH INSULATED
COVER

BUBBLE
WINDOW

MAXIMUM INPUT BUBBLE WINDOW WITH
MOVABLE INSULATION PANELS

CORRUGATED METAL

PAINTED STEEL SLID DOME

SPRAY FOAM AND
FIREPROOF COVER

HAND WINCH TO OPEN,
CLOSE AND ADJUST
ANGLE OF INSULATED
REFLECTOR PANELS

PAINTED, CORRUGATED
STEEL SIDING
WOOD INSIDE
FIBERGLASS ROLL
INSULATION INBETWEEN

SECTION

139

East elevation

North entrance

South facing deck

140

Southwest elevation

Sliding insulated shutter

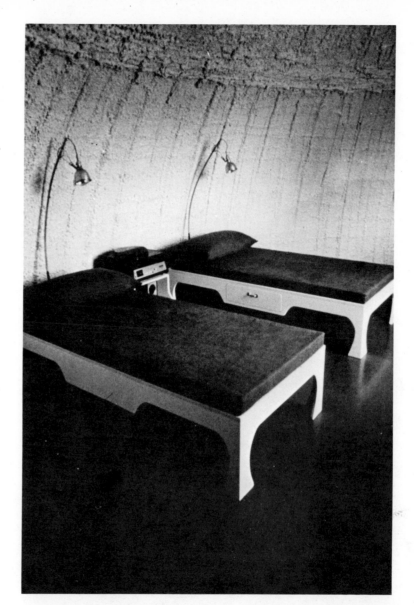

Spherical shape wall   Photos by Michael Jantzen

141

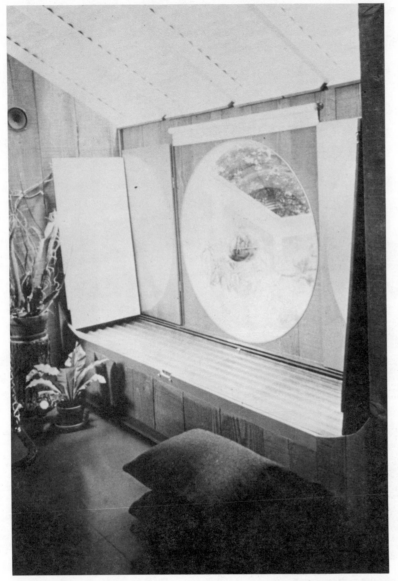

Storage tank

Photo by Michael Jantzen

THE JANTZEN HOUSE II

LOCATION: Carlyle, Illinois.
LATITUDE: 38.5 degrees N.
REGION: Temperate.
DESIGNER, SOLAR ENGINEER AND BUILDER: Michael
    Jantzen, Carlyle, Illinois.

SPACIAL SOLUTION

This house is located on a 17-acre, heavily wooded and animal populated site, some 50 miles east of St. Louis. It sits on top of a ridge and its longest side faces south. Gardens of special herbs with rodent and insect repellent properties are planted around the house, and there are vents near the gardens to allow the herbs' fragrances to freshen the inside air. Bird feeders are located at various points around the house to encourage the presence of birds and their contribution in controlling the insect population. The several green houses placed throughout the site provide 70 percent of the food consumed by the Jantzens.

The house was erected on a concrete foundation wall, with the first floor raised 3 feet above the ground level. It is constructed of wood framed, laminated arches that are covered with corrugated steel and insulated with 3 to 4 inches of sprayed foam. All of the vertical walls are covered with plywood, with pebble aggregate glued to the surface. The bubble windows located on the south facade are equipped with sliding or hinged, insulating, reflective shutters. Operable air vents were placed beneath each window. The house has many operable louvers, vents, vent covers, plugs, partitions and other devices which permit any of the rooms to be heated independently from the others.

Water is pumped from a well and is stored in a 120-gallon tank located on the second floor. The water flows by gravity, which supplies enough force for all the high pressure spray nozzles, and reduces the amount of water consumed by increasing the efficiency of the spray. A "pure-way" bio-flow toilet, which operates by dropping a package of yeast culture into it once a week to accelerate decomposition, is used. Fluorescent lighting was installed in areas that make use of prolonged artifical lighting. Incandescent lighting was installed in areas that make limited use of artifical lighting. Over-all lighting requirements are low, due to the bright white surfaces in the interior spaces. The Jantzens also use a special full-size electric refrigerator which consumes the same amount of energy as a 25-watt light bulb.

MECHANICAL SOLUTION

The 2,372 square feet of floor area are heated by three separate heating systems. The main system is a passive solar unit consisting of 15 tons of stones located in the south end of the house, directly below the living room. The storage bin is capable of heating the entire house. The secondary system consists of 320 square feet of air-cooled collectors. The third system is a wood burning stove which can heat the entire house with a few logs. There are operable louvered vents above the stove to direct the heat into the bedroom.

The domestic water is heated by 64 square feet of flat-plate collectors located below the living room window. Heating elements are located in line near each faucet to provide the necessary energy to raise the water temperature. This set-up greatly reduces the heat losses that take place when the water travels through the pipes.

The spacial and mechanical concept implemented in this house provides 100 percent of the space heating and cooling needs and 80 percent of the domestic hot water.

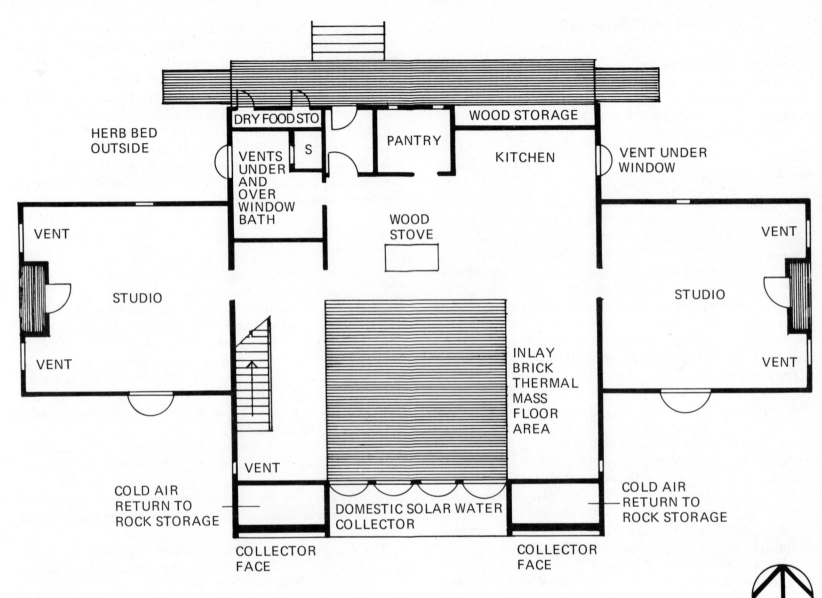

HERB BED
OUTSIDE

DRY FOOD STO

VENTS
UNDER
AND
OVER
WINDOW
BATH

S

PANTRY

WOOD STORAGE

KITCHEN

VENT UNDER
WINDOW

WOOD
STOVE

VENT

VENT

STUDIO

VENT

STUDIO

VENT

INLAY
BRICK
THERMAL
MASS
FLOOR
AREA

VENT

COLD AIR
RETURN TO
ROCK STORAGE

DOMESTIC SOLAR WATER
COLLECTOR

COLD AIR
RETURN TO
ROCK STORAGE

COLLECTOR
FACE

COLLECTOR
FACE

FLOOR PLAN

0 1   5   10 FEET

SCALE

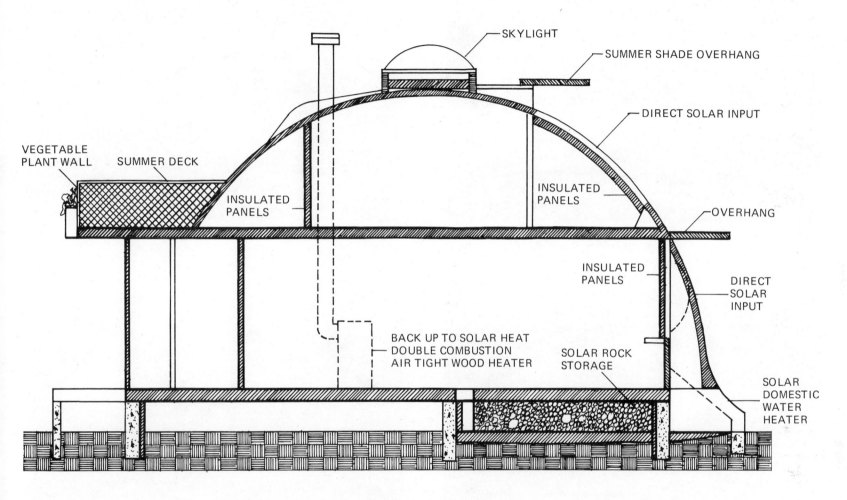

SKYLIGHT

SUMMER SHADE OVERHANG

DIRECT SOLAR INPUT

VEGETABLE
PLANT WALL

SUMMER DECK

INSULATED
PANELS

INSULATED
PANELS

OVERHANG

INSULATED
PANELS

DIRECT
SOLAR
INPUT

BACK UP TO SOLAR HEAT
DOUBLE COMBUSTION
AIR TIGHT WOOD HEATER

SOLAR ROCK
STORAGE

SOLAR
DOMESTIC
WATER
HEATER

CROSS-SECTION

145

South elevation

Northwest elevation

Southwest elevation

146

Bedroom

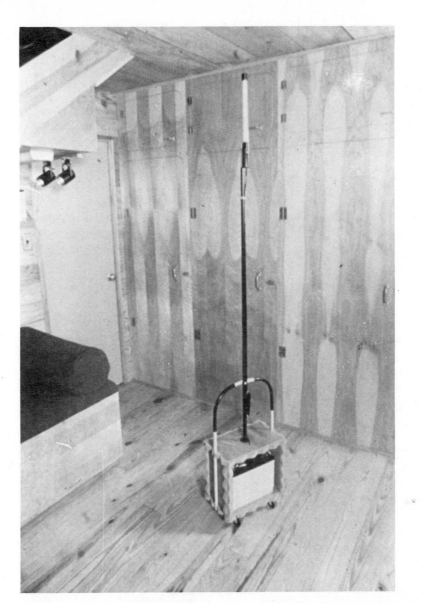

Livingroom

Closets

Photos by Michael Jantzen

147

THE THOMASON SOLAR HOUSE NO. 6

LOCATION: Malboro, Maryland.
LATITUDE: 39 degrees N.
REGION: Temperate.
DESIGNER AND BUILDER: Thomason Solar Homes, Inc., Oxon Hills, Maryland.

SPACIAL SOLUTION

This house has two stories and a basement. The first story has a heated area of 1,550 square feet and the basement has 1,400 square feet. The second floor is not directly heated. The house is rectangular in plan and faces 15 degrees west of south. The wood frame construction uses 3-1/2 inches of fiberglass insulation for the walls and 6 inches for the ceiling. The interior layout uses closets, vestibules and storage areas to create buffer zones that effectively increase the over-all insulation. Windows on the first floor are double glazed; on the second floor, they are single glazed.

MECHANICAL SOLUTION

There are 960 square feet of flat-plate, water-cooled collectors mounted on the roof at a tilt of 45 degrees. Each collector is composed of a single cover plate and a corrugated aluminum sheet. The sheet, which is coated with black paint, acts as the absorber plate. Filtered water is allowed to trickle down the trough of the absorber and is collected at the bottom of the panel in a manifold. From there, the water is circulated to the 1,600-gallon water storage tank, which is imbedded in 28 tons of stones. As the water in the tank is heated, the rocks absorb their thermal energy. Heat is distributed to the rooms by a 1/4-horsepower blower located at the bottom of the bin. The house does not use supply or return ducts; instead, it uses the air space between the basement ceiling and the first floor to channel the air.

SOLAR SYSTEM SCHEMATIC

North elevation

Southwest elevation

Domestic hot water is preheated in a 40-gallon tank contained in the main water storage tank. The final energy boost is provided by a separate 40-gallon, oil fired water heater.

Directly over the stones in the storage bin, there is an array of copper tubes, connected in a closed loop to the water heater. When additional heat is needed and the stone storage tank does not have the capacity to supply it, the water heater circulates warm water through the tubes. As the blower circulates the air flow upward, the tubes heat the air that was delivered to the rooms. Space cooling is provided by a conventional air conditioner operated at night to cool the rocks. During the day, air is blown from the rooms through the bin, where it is cooled, and then back to the rooms. The air conditioning unit is operated at night to make use of the off-peak rates.

THE ZWILLINGER HOUSE

LOCATION: New Boston, New Hampshire.
LATITUDE: 43 degrees N.
REGION: Temperate.
SOLAR ENGINEER, CONTRACTOR AND OWNER: D.
    Zwillinger, New Boston, New Hampshire.

SPACIAL SOLUTION

This 3,000-square-foot, two-and-a-half-story house has a built-in garage, a basement and a root cellar. The wood framed walls are insulated with 2 inches of polyurethane. The windows are small in size and are double glazed.

MECHANICAL SOLUTION

The south elevation contains 1,300 square feet of air-cooled, flat-plate collectors that form an integral part of the structural elements of the house. The wall and collectors are tilted to 57 degrees from the horizontal. Air is circulated through the collectors and the 200 tons of stones in the storage tank. The rooms can be heated with air blown directly from the collectors or from the storage tank. No summer cooling is needed.

The solar system provides 90 percent of the space heating needs, with the remainder supplied by a wood burning stove. A full capacity electrical heating system was also installed.

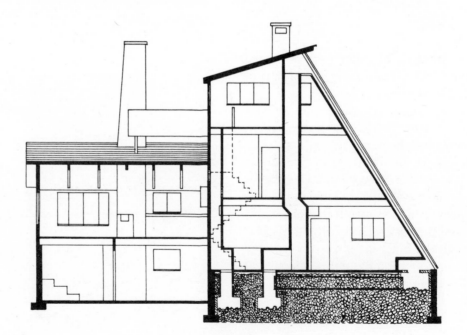

CROSS SECTION

150

FIRST FLOOR PLAN

SECOND FLOOR PLAN

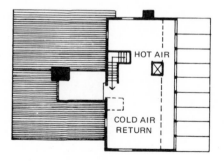

THIRD FLOOR PLAN

West elevation

Southeast elevation

151

THE KELBAUGH HOUSE

LOCATION: Princeton, New Jersey.
LATITUDE: 40 degrees N.
REGION: Temperate.
ARCHITECT AND SOLAR ENGINEER: Douglas Kelbaugh,
   Princeton, New Jersey.

SPACIAL AND MECHANICAL SOLUTION

This 2,100-square-foot, two-story house uses a passive solar heating system. The house sits on the north side of the 60-foot by 100-foot lot, avoiding the shadows cast by the neighboring houses. The wood framed walls and ceilings are insulated with newspaper pulp, since this material has the highest insulating properties for the lowest cost. The walls achieve a U factor of 0.05 and the ceilings achieve a U factor of 0.02. There are few windows in the house and insulating glass was installed in all of them.

The exterior envelope is covered with rough-sewn cedar plywood. The south wall constitutes the collector for the solar system, which has an area of 800 square feet and was built with 15-inch thick concrete blocks. The blocks were painted black and covered with two sheets of glass that act as the cover plates. When the sun warms the wall, the air around it rises and is channeled through vents into the rooms. This air flow pulls cool air from the bottom of the rooms into the wall, where it is heated once again. A green house located on the southeast corner of the house provides additional heat.

The house is cooled in the summer by natural and forced ventilation. The circulation of night air cools the wall, which, in turn, cools the room during the day. The wall is shaded by two deciduous trees, which prevent heat collection during the summer. The coupling of the solar system and the spacial concepts provides 75 percent of the space heating needs. A gas fired, forced air furnace heats the rooms via ducts imbedded in the south wall.

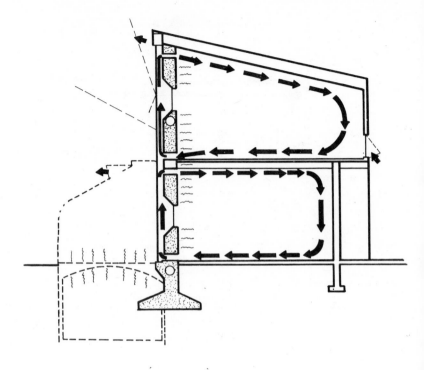

SOLAR HEATING SYSTEM

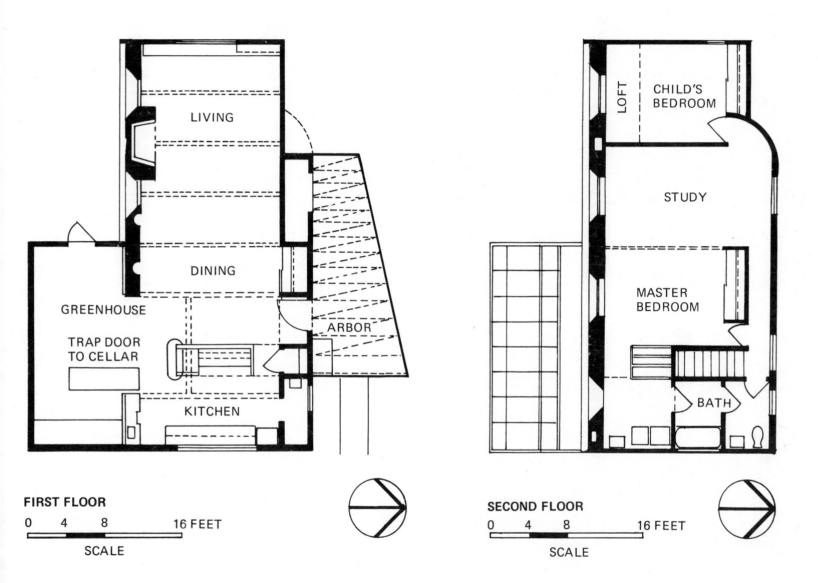

FIRST FLOOR

LIVING

DINING

GREENHOUSE

TRAP DOOR
TO CELLAR

ARBOR

KITCHEN

0   4   8   16 FEET

SCALE

SECOND FLOOR

LOFT

CHILD'S
BEDROOM

STUDY

MASTER
BEDROOM

BATH

0   4   8   16 FEET

SCALE

Southwest elevation

South, facing greenhouse　　　　　**Photos by John Coursen**

THE EVANS HOUSE

LOCATION: East Hampton, New York.
LATITUDE: 41 degrees N.
REGION: Temperate.
CONSULTING ENGINEER: B. Anderson, Harrisville, New
　.Hampshire.

SPACIAL AND MECHANICAL SOLUTION

This 2,100-square-foot house has one story and a loft area. Its wooden structure is erected in an A-frame configuration and is well-insulated with ureaformaldehyde foam. The south wall contains 750 square feet of air-cooled double glazed panels, tilted to 60 degrees from the horizontal. The panels allow solar radiation to penetrate the interior space where the thermal energy is stored in the walls and the floor. A fan circulates the air from the room through the storage element, which consists of 1,000 1-gallon containers filled with water and enclosed by an insulated concrete bin. A 5-ton heat pump is used in combination with the heat storage element. The domestic hot water is heated by two flat-plate collectors mounted on the south wall. The combination of thermal storage and the heat pump provides 100 percent of the space heating needs.

Interior

East elevation

West elevation

THE WENNING HOUSE

LOCATION: La Grangeville, New York.
LATITUDE: 42 Degrees N.
REGION: Temperate.
ARCHITECT: Wenning Associates, La Grangeville, New York.
SYSTEM MONITOR: General Electric Company.

SPACIAL SOLUTION

This two-story, 3,200-square-foot house has four bedrooms, a full basement and a two-car attached garage. The design incorporates south-facing windows with louvered overhangs to provide summer shade. A closed water loop conserves energy from the fireplace, kitchen appliances and laundry and bathroom fixtures.

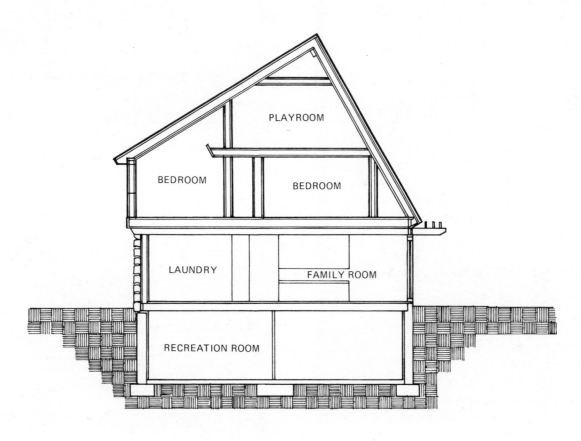

SECTION

0   4   8   16 FEET
SCALE

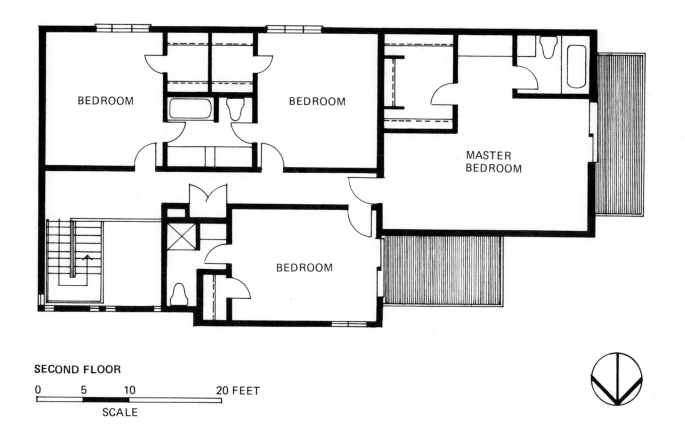

SECOND FLOOR

```
0      5    10              20 FEET
|======|====|=======|========|
              SCALE
```

MECHANICAL SOLUTION

The 1,200 square feet of water-cooled, flat-plate collectors are integrated into the roof structure, which supports them at a 60 degree angle from the horizontal. Water is circulated through the collectors to the 4,000-gallon storage tank located under the garage slab. Heat is delivered to the rooms by a three-zone forced air system. Two General Electric air-to-air heat pumps serve as auxiliary back-up. The pumps are operated in the summer to provide space cooling. The combined system is predicted to supply 70 percent of the space heating and 80 percent of the water heating.

157

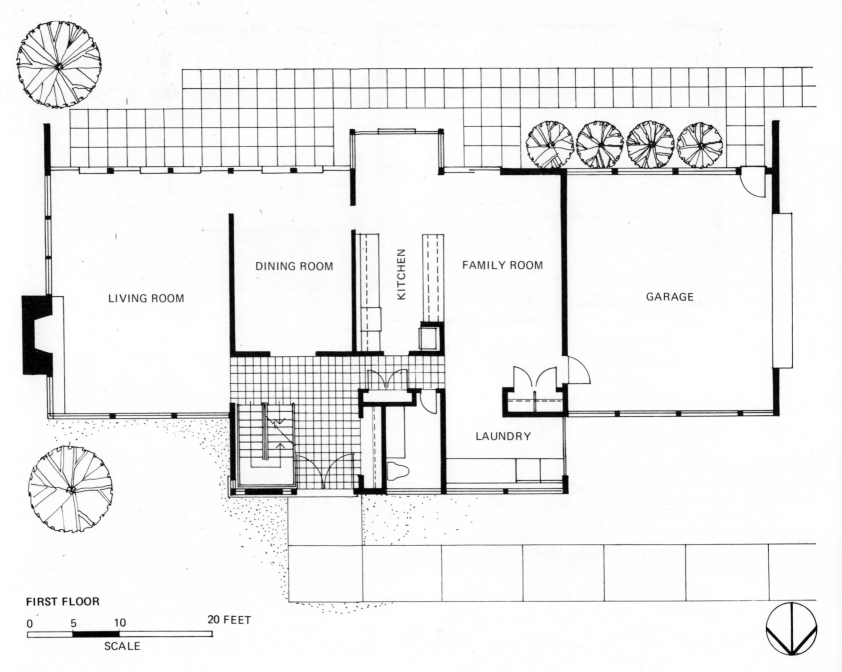

LIVING ROOM

DINING ROOM

KITCHEN

FAMILY ROOM

GARAGE

LAUNDRY

FIRST FLOOR

0 5 10 20 FEET

SCALE

THE BARBASH HOUSE

LOCATION: Quoque, New York.
LATITUDE: 41 degrees N.
REGION: Temperate.
ARCHITECT: J. S. Whedbee, New York, New York.
SOLAR ENGINEER: Owens-Illinois, Chicago, Illinois.

SPACIAL SOLUTION

This wood frame two-story house has 3,000 square feet of floor area. The first level houses a car port, a north-facing studio and a utility room. The kitchen, living room, two bedrooms and a medium size green house are on the second floor. Wooden balconies and sundecks surround most of the second floor. The north- and south-facing walls are tilted to 32.5 degrees and 57.5 degrees from the horizontal, respectively. The walls and ceiling have 6 inches of fiberglass insulation.

South elevation

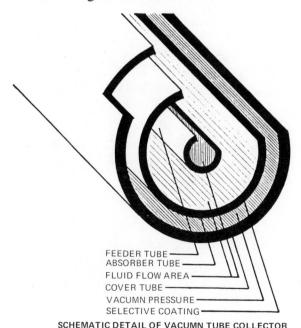

FEEDER TUBE
ABSORBER TUBE
FLUID FLOW AREA
COVER TUBE
VACUMN PRESSURE
SELECTIVE COATING
SCHEMATIC DETAIL OF VACUMN TUBE COLLECTOR

MECHANICAL SOLUTION

The solar heating system provides space and domestic water heating. The collector is composed of 510 square feet of vacuum-jacketed, liquid-cooled glass tubes, tilted to 57.5 degrees from the horizontal. This type of collector reduces convective and conductive heat losses, thereby eliminating the need for an automatic draining valve or for the use of antifreeze.

Water circulates from the collectors to the 1,000-gallon storage tank. Heat is distributed to the rooms by fan-coil units in the forced air system. No space cooling is provided.

The solar system provides 80 percent of the energy needs, and two fireplaces, each equipped with grates, circulate water from the storage tank to provide the make-up energy.

159

West elevation

Southeast elevation

Collector detail

160

THE ENGLE HOUSE

LOCATION: Wagoner, Oklahoma.
LATITUDE: 35 degree N.
REGION: Temperate.
ARCHITECT AND SOLAR ENGINEER: Alan Lower and
  Associates, Oklahoma.

SPACIAL SOLUTION

The 3,700 square feet of floor area of this house are divided into four interlocking levels. The exterior is of stone and wood shingles with earth berms for additional protection against heat losses. All the glass area is double glazed, with deep overhangs on the south face adding protection against the summer sun. A massive stone fireplace, which adds to the internal thermal mass, is a focal point for the interior spacial layout.

MECHANICAL SOLUTION

The solar system is composed of 760 square feet of flat-plate, water-cooled collectors, mounted on the south-facing roof, which is tilted to 50 degrees from the horizontal. The collectors are equipped with an automatic draining valve. A 1/3-horsepower pump circulates water through the collector and the 2,000-gallon concrete storage tank. Heat is delivered to the room by fan-coil units. Cooling is provided by conventional means.

The system provides 50 percent of the space heating and 100 percent of the domestic water heating. The make-up energy is provided by an electric boiler.

Southeast elevation

Southwest elevation

THE BISHOPRICK HOUSE

LOCATION: Salem, Oregon.
LATITUDE: 45 degrees N.
REGION: Temperate.
ARCHITECT AND SOLAR ENGINEER: W. Bishoprick. Salem, Oregon.

SPACIAL SOLUTION

This house consists of 1,450 square feet of floor area and is located on a steeply sloping, densely wooded site. There are three levels, with two bedrooms, two baths, a living room, a dining room, a kitchen and a carport. The exterior walls are wood shingled and insulated with batt fiberglass to yield a total resistance value of $R = 11$. The ceiling is also wood shingled, with 6-inch batt insulation in the built-up roof yielding a total resistance value of $R = 19$. There are 315 square feet of glass area (double glazed).

MECHANICAL SOLUTION

The heat collection element consists of 460 square feet of home designed, air-cooled, flat-plate collectors, tilted to 60 degrees from the horizontal. The collector was constructed by placing a 3-foot wide section of corrugated aluminum siding 3 inches below the double glazed cover plate. A 3/4-horsepower fan blows the air through the collector and the 50 tons of stones that comprise the storage element.

Heat is delivered to the rooms by a 1/4-horsepower blower. Electric resistance elements located in the duct system provide back-up energy when needed. Cooling is provided by circulating room air— which has been cooled during the night by blowing through the storage bin.

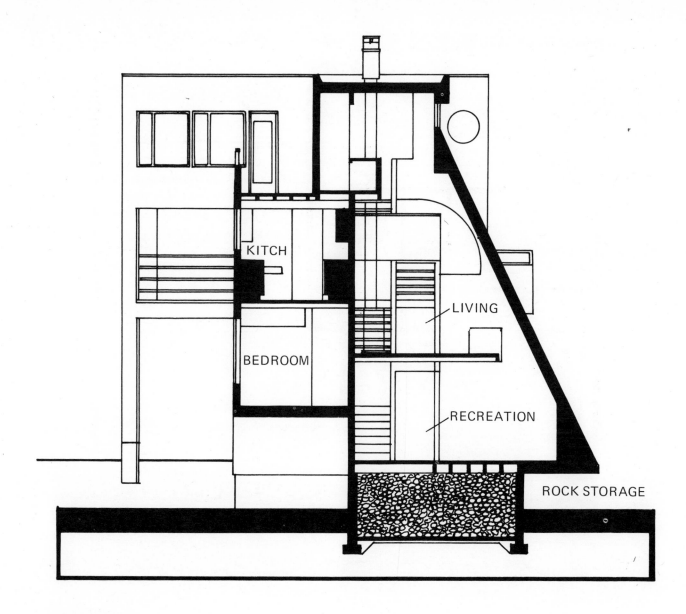

KITCH

LIVING

BEDROOM

RECREATION

ROCK STORAGE

SECTION A-A

0   5 FEET

SCALE

163

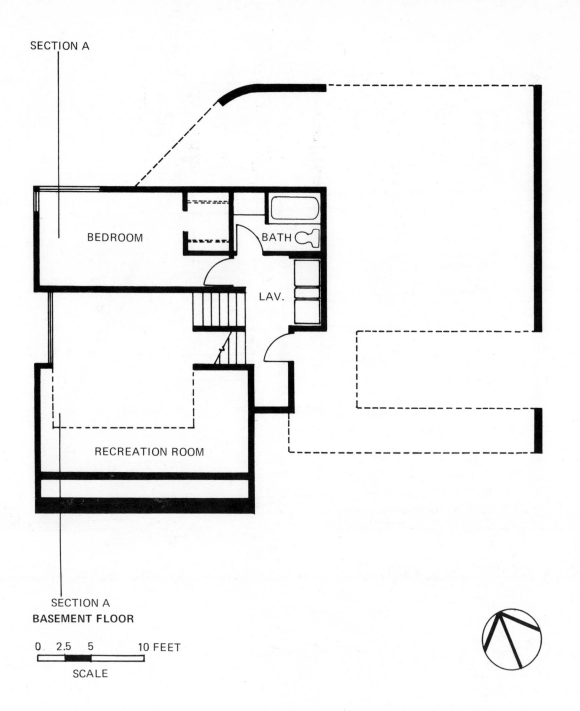

SECTION A

BEDROOM

BATH

LAV.

RECREATION ROOM

SECTION A
BASEMENT FLOOR

0   2.5   5   10 FEET
SCALE

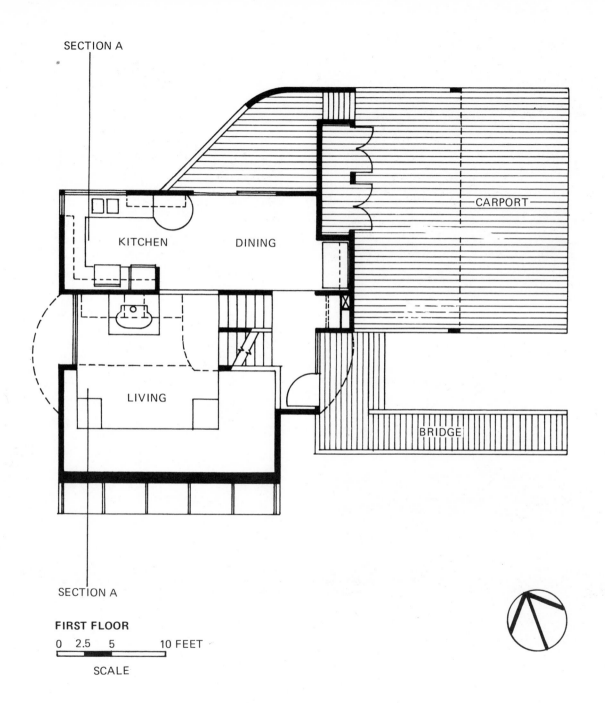

SECTION A

KITCHEN   DINING

CARPORT

LIVING

BRIDGE

SECTION A

FIRST FLOOR

0   2.5   5   10 FEET

SCALE

South elevation

Southeast elevation

Collector detail

166

THE PEOPLE SPACE COMPANY HOUSE

LOCATION: Windham, Vermont.
LATITUDE: 43 degrees N.
REGION: Temperate.
ARCHITECT AND SOLAR ENGINEER: People Space Company,
    Robert Shannon, Windham, Vermont.

SPACIAL AND MECHANICAL SOLUTION

This house is used as a ski cottage and has 900 square feet of floor area divided into two floors and a half-basement. Its form closely resembles that of a cube. The south wall contains 400 square feet of air-cooled, flat-plate collectors, tilted to 60 degrees and 90 degrees from the horizontal. The collectors are also equipped with two reflector planes, each one measuring 8 feet by 20 feet. These reflectors are adjustable and can be closed at night to further insulate the house. During the summer months,

Northwest elevation

South elevation

167

the top reflector is adjusted to block the sun. A 1-horsepower blower circulates air through the collectors to the 20 tons of stone in the storage bin and to the rooms. The domestic water is preheated by the collectors. There are no cooling requirements.

The solar system provides 70 percent of the space heating requirements, with the make-up portion being supplied by the fireplace and electric heaters.

Southeast elevation

Section 7. Hot Arid Region

The buildings presented in this section are the following:

The Frerking House. Chino Valley, Tucson, Arizona.

The Keating House. Cornville, Arizona.

The Sandia Heights House. Albuquerque, New Mexico.

The Baer House. Corrales, New Mexico.

The Suncave House. Sante Fe, New Mexico.

The Terry House. Sante Fe, New Mexico.

THE FRERKING HOUSE

LOCATION: Chino Valley, Arizona.
LATITUDE: 34 degrees N.
REGION: Hot Arid.
ARCHITECTS AND DESIGNERS: M. Frerking and W. Otwell, Chino Valley, Arizona.

SPACIAL SOLUTION

This house consists of 550 square feet of floor area with an additional 150 square feet of greenhouse area. The walls are made of 12-inch thick adobe block, the outside of which is insulated with an inch of styrofoam that is protected with a stucco layer. The ceiling has a wood frame with metal roofing and 6 inches of insulation.

The greenhouse acts as the collector, allowing the winter sun to penetrate into the house and be stored in the walls and floor slab. In the summer, the greenhouse is covered with a wooden snow fence to provide shade. The domestic water is heated by a flat-plate collector which is detached from the house and located on the west end. This passive solar system provides over 80 percent of the space heating needs. The make-up energy is provided by a wood stove.

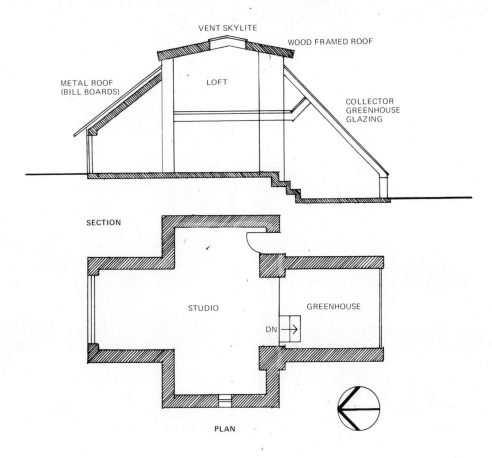

171

Southwest elevation

Northeast elevation

THE KEATING HOUSE

LOCATION: Cornville, Arizona.
LATITUDE: 34 degreees N.
REGION: Hot Arid.
ARCHITECT AND SOLAR ENGINEER: Low Energy Design,
    Tucson, Arizona.

SPACIAL AND MECHANICAL SOLUTION

This house has 1,500 square feet of floor area divided among four stories. The first floor is partially below grade and houses the main storage tank. The second floor consists of the living room, dining room and kitchen. Four bedrooms are on the third and fourth floors.

There are 400 square feet of flat-plate, water-cooled collectors on the roof of the fourth floor. The heated water is stored in a 1,000-gallon capacity tank imbedded in 30 tons of stone. Heat is distributed to the rooms by a forced air system. Domestic water is heated by a special panel located in the middle of the roof and its storage tank is located on the fourth floor.

Southeast elevation

Northwest elevation

THE SANDIA HEIGHTS HOUSE

LOCATION: Albuquerque, New Mexico.
LATITUDE: 35 degrees N.
REGION: Hot Arid.
ARCHITECTS: Burns and Peters, Albuquerque, New Mexico.
SOLAR ENGINEER AND PRODUCT MANUFACTURER:
   Albuquerque Western Solar Industries, Albuquerque, New Mexico.
BUILDER: Sandia Homes, Inc.

SPACIAL SOLUTION

This wood frame house has 2,200 square feet of floor area divided into two floors; in addition, it has a two-vehicle carport. The walls are insulated with 4 inches of fiberglass and the ceiling with 10 inches of fiberglass. All windows are double glazed and the south-facing windows are recessed to prevent the summer sun from entering the house.

MECHANICAL SOLUTION

The solar system provides space and domestic water heating. There are 640 square feet of concentrating, tracking, water-cooled collectors, divided into two 60-foot rows and tilted to 45 degrees from the horizontal. The absorbing tube is stationary, while the reflector rotates to provide maximum optimization of the sun's rays. The collectors are automatically draining when sun tracking is interrupted.

A 1-horsepower pump circulates the water through the collectors and to the 1,000-gallon capacity storage tank. Heat is delivered to the room by fan-coil units in the forced air system. Domestic hot water is preheated in a 40-gallon tank contained in the main storage tank. Additional heat is provided when needed by an electric hot water heater tank. Evaporative cooling is used during the summer.

The solar system provides 75 percent of the space heating needs. The make-up energy is supplied by electric heating elements in the duct system.

173

South elevation

East elevation

174

THE BAER HOUSE

LOCATION: Corrales, New Mexico.
LATITUDE: 35 degrees N.
REGION: Hot Arid.
DESIGNER AND OWNER: Steven Baer, Zomeworks Inc.,
    Albuquerque, New Mexico

SPACIAL AND MECHANICAL SOLUTION

This one-story house is located on a 4-acre site. It is composed of eleven zoned volumes arranged in a "V"-shaped floor plan. Portions of the walls are built with adobe to increase the thermal storage. The walls and ceiling are insulated extremely well, with the exception of the southern wall. Glass areas in the house are kept to a minimum. Each of the four south walls has an approximate area of 10 feet by 10 feet and is covered with single pane glass. Behind each area, there is a stack of oil drums painted black and filled with water. In front of the glass, there is a hinged, aluminum faced, insulated wall, which is open during the winter days to allow the sun to heat the oil drums. At night, the wall is rotated into a closed position to prevent heat loss. Cooling is provided during the summer by allowing the barrels to radiate the heat to the exterior at night and closing the walls during the days.

Water is pumped out of the ground by a windmill and it flows into the house by gravity. Domestic water is heated by a detached solar collector.

This passive system provides 80 percent of the space heating needs, with the make-up energy supplied by several wood burning stoves.

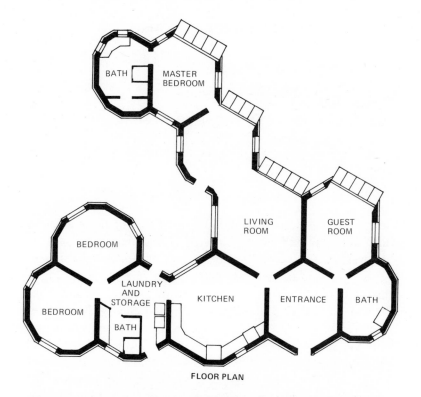

FLOOR PLAN

Back Patio

175

THE SUNCAVE HOUSE

LOCATION: Sante Fe, New Mexico.
LATITUDE: 35.5 degrees N.
REGION: Hot Arid.
ARCHITECT AND SOLAR ENGINEER: David Wright, Sea Ranch, California.
SOLAR CONSULTANT: M. Chalom, Santa Fe, New Mexico

SPACIAL AND MECHANICAL SOLUTION

This split-level house has 1,350 square feet of floor area. The walls and floor slab are constructed of adobe and concrete. The east, west and north walls and the roof are insulated with earth berms. The south wall has 440 square feet of vertical glass, which allows the winter sun to penetrate into the interior space. The walls and floor slab store the thermal energy. At night, the glass area is covered with insulated shutters. An overhang prevents the summer solar rays from entering into the house. The domestic water is preheated by a solar unit. This passive solar system provides over 85 percent of the annual space heating needs, with the remaining percentage supplied by two fireplaces and small electric baseboard heaters.

Southwest elevation

Southeast elevation　　　　　　**Photos by David Wright**

176

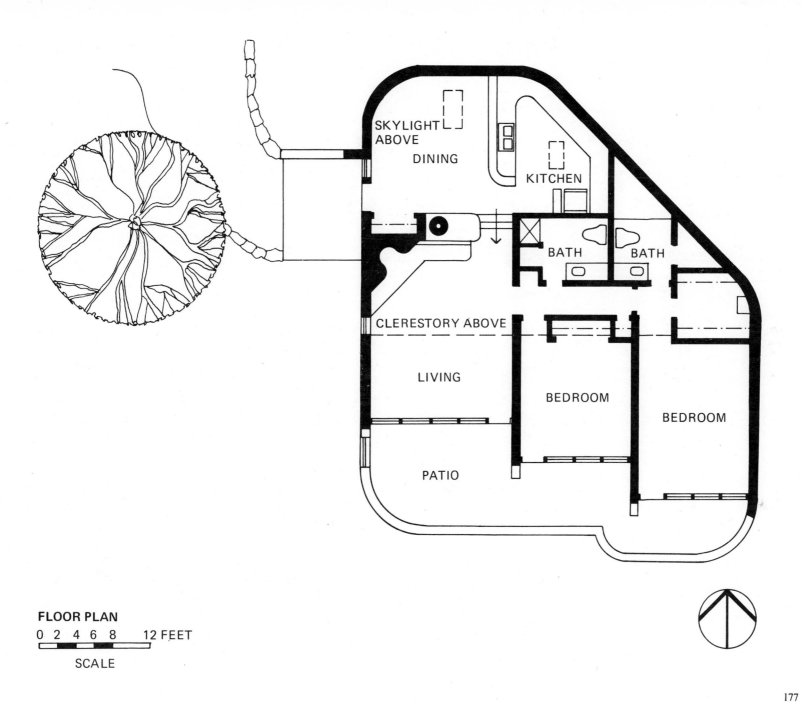

SKYLIGHT
ABOVE

DINING

KITCHEN

BATH

BATH

CLERESTORY ABOVE

LIVING

BEDROOM

BEDROOM

PATIO

FLOOR PLAN

0 2 4 6 8   12 FEET

SCALE

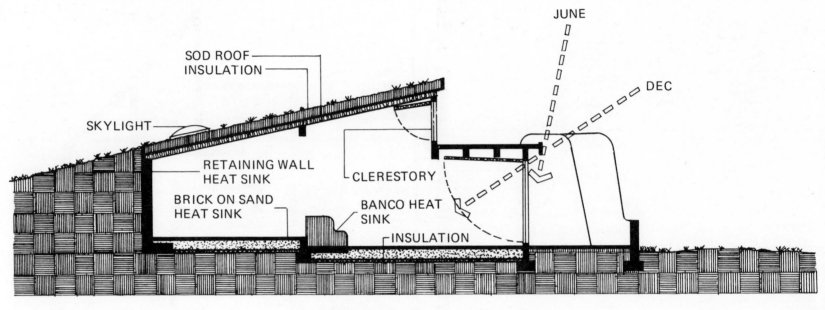

SOD ROOF
INSULATION

SKYLIGHT

RETAINING WALL
HEAT SINK

BRICK ON SAND
HEAT SINK

CLERESTORY

BANCO HEAT
SINK

INSULATION

JUNE

DEC

SECTION

0 2 4 6 8   12 FEET

SCALE

178

Dining Room   **Photo by David Wright**

THE TERRY HOUSE

LOCATION: Santa Fe, New Mexico.
LATITUDE: 35 degrees N.
REGION: Hot Arid.
DESIGNER: Karen Terry, Santa Fe, New Mexico.
ARCHITECT AND SOLAR ENGINEER: David Wright, Sea
    Ranch, California.

SPACIAL AND MECHANICAL SOLUTION

This four-level house has 850 square feet of floor area, and each interior function is located on the level best suited to the heat comfort demand. The work studio is on the lowest level because of its low heat gain; living and cooking areas are on the second level; the bath on the third level; and sleeping is on the top level, for large heat gain and stratification.

The walls are constructed of stucco-covered adobe and act as heat storage areas. In addition, 1,100 gallons of containerized water are buried inside the interior walls for extra heat storage. The sloping, south-facing glass wall permits the winter sun to penetrate the interior space where the walls and floor slab store the thermal energy. The wall is covered with wooden louvers that keep the summer sun out while creating shadow patterns. Natural ventilation is used for space cooling. Domestic water is heated by solar collectors located east of the house.

This passive solar system provides 95 percent of the space heating needs, with the small make-up percentage supplied by wood burning stoves.

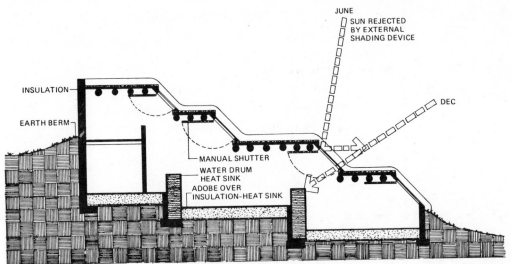

JUNE
SUN REJECTED
BY EXTERNAL
SHADING DEVICE

DEC

INSULATION

EARTH BERM

MANUAL SHUTTER
WATER DRUM
HEAT SINK
ADOBE OVER
INSULATION-HEAT SINK

SECTION

0 2   5   10
SCALE

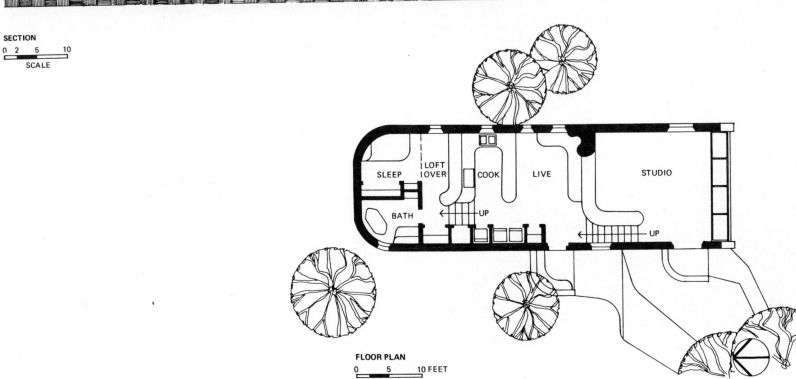

SLEEP | LOFT OVER | COOK | LIVE | STUDIO

BATH

UP

UP

FLOOR PLAN

0   5   10 FEET
SCALE

180

Main entrance

Interior

Southwest elevation   **Photos by David Wright**

Section 8. Hot Humid Region

The buildings presented in this section are the following:

The Delap House. Fayetteville, Arkansas.

The Kim House. Fayetteville, Arkansas.

The Austin House. Austin, Texas.

THE DELAP HOUSE

LOCATION: Fayetteville, Arkansas.
LATITUDE: 36 degrees N.
REGION: Hot Humid.
ARCHITECT: James Lambeth, Fayetteville, Arkansas.
CONSULTING ENGINEER: J. Wall, University of Arkansas.

SPACIAL AND MECHANICAL SOLUTION

This 2,000-square-foot three-level house rests on a wooded and south-sloping 1-acre site. The house is in the shape of a fan, with the north wall the short end. The east and west walls are windowless and angle out to meet the south wall. The south facade is composed of a large glass area that covers 860 square feet of a block wall that is painted black. The 12-inch thick black wall acts as the collector and a portion of the stored heat is conducted into the interior spaces by convection. The majority of the collected heat is stored in rocks located in the crawl space below grade. A forced air system distributes this heat to the rooms. When the temperature in the storage bin is inadequate, a heat pump is used to heat the rooms. In addition to this, a fireplace also provides back-up heat.

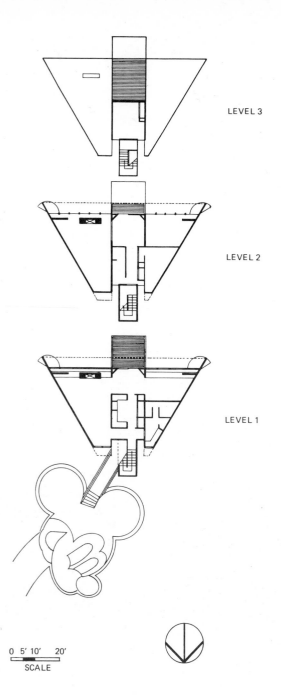

LEVEL 3

LEVEL 2

LEVEL 1

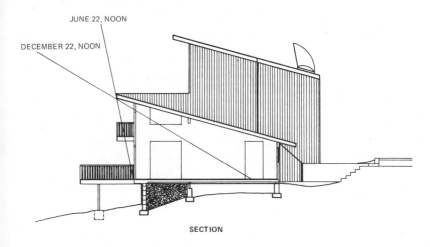

JUNE 22, NOON

DECEMBER 22, NOON

SECTION

0 5' 10'   20'
SCALE

"Mickey Mouse" parking space

South wall collector

Northwest elevation

186

THE KIM HOUSE

LOCATION: Fayetteville, Arkansas.
LATITUDE: 36 degrees N.
REGION: Hot Humid.
ARCHITECT AND SOLAR DESIGNER: James Lambeth,
    Fayetteville, Arkansas.

SPACIAL AND MECHANICAL SOLUTION

There is no conceptual difference between this house and the
Delap house. However, there are spacial and mechanical details
incorporated in the Kim house that reflect a refinement over
those used in the Delap house. For example, the west wall re-
sponds to the winter sun and westerly views. In addition, another 75 square feet of water-cooled, flat-plate collectors were
incorporated into the roof to preheat the domestic hot water. A
large fireplace in the northeast wall supplies part of the make-up
heat.

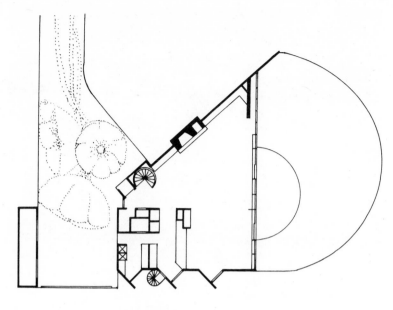

LEVEL 1

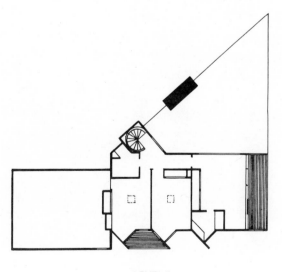

LEVEL 2

South wall collectors

Photo by James Lambeth

THE AUSTIN HOUSE

LOCATION: Austin, Texas.
LATITUDE: 30 degrees N.
REGION: Hot Humid.
ARCHITECT: Thomas Leach and Associates, Austin, Texas.
ENGINEER: Hammer, Inc., Austin, Texas.

SPACIAL SOLUTION

This large, contemporary house has 4,050 square feet of floor area, organized into a compact three-story plan. The house is situated on the coolest portion of the site to make maximum use of the cool winds and shades. The roof planes were designed to shelter the entrance and to funnel the prevailing winds.

MECHANICAL SOLUTION

The solar system provides space heating and cooling and domestic hot water. The collector is composed of 60 liquid-cooled linear concentrating units, detached from the house to increase solar absorption. The storage tank has a capacity of 1,800 gallons and is equipped with a heat exchanger. The heated water is circulated from the tank to fan-coil units in the forced air system. Electric resistance coils supply the back-up energy. Cooling is provided by an absorption chiller which is assisted by the solar system. The domestic hot water is preheated by a heat exchanger in the main storage tank and a conventional water heater provides the back-up when necessary.

PART III
COMMERCIAL BUILDINGS

Section 9. Cool Region

The buildings presented in this section are the following:

Architects Office Building. Detroit, Michigan.

Home Savings & Loan Association. Stoughton, Wisconsin.

ARCHITECTS OFFICE BUILDING

LOCATION: Detroit, Michigan.
LATITUDE: 42.5 degrees N.
REGION: Cool.
ARCHITECT: Smith, Hichman & Grylls, Inc., Detroit, Michigan.

MECHANICAL SOLUTION

This retrofit project was designed to test the feasibility of a solar energy system in northern latitudes and, specifically, in urban areas, where air pollution and adjacent structures may have a detrimental impact on the system.

The building uses 10,000 square feet of tubular glass, vacuum-jacketed, liquid-cooled collectors, mounted on an adjustable tilt frame. The advantage of the vacuum tube is that it reduces night-time heat losses, thereby eliminating the need for insulation and draining valves. Broken tubes may be replaced manually. The collector and the 1,500-gallon storage tank provide domestic hot water, heat for the baseboard radiators and power for the absorption chiller and the cooling tower basin.

The system conditions 25,000 square feet of prime office space.

The project was successful in proving the "technical" feasibility of solar energy systems under the above-mentioned conditions.

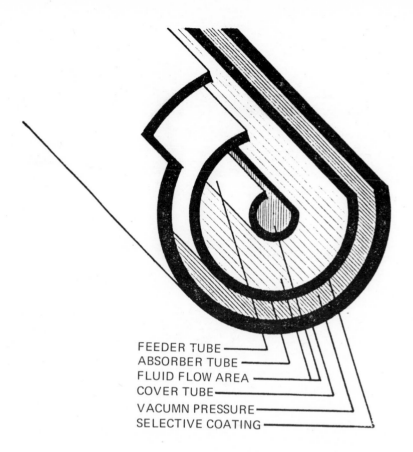

FEEDER TUBE
ABSORBER TUBE
FLUID FLOW AREA
COVER TUBE
VACUMN PRESSURE
SELECTIVE COATING

SCHEMATIC DETAIL OF VACUMN TUBE COLLECTOR

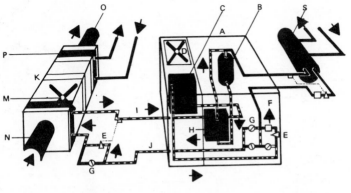

Cooling
A. Heat pump
  B. Compressor
  C. Condenser coil
  D. Condenser fan
  F. Solenoid valve
  G. Typical check valve
  H. Diverter valve
  I. Low pressure cool gas
  J. High pressure warm liquid
K. Air handling unit
  L. Heating coil
  M. Fan
  N. Air from space
  O. Air to space
  E. Expansion valve
  G. Check valve

Normal Heating
A. Heat pump
  B. Compressor
  C. Condenser coil
  D. Condenser fan
  E. Expansion valve
  F. Solenoid valve
  G. Typical check valve
  H. Diverter valve
  I. High pressure hot gas
  J. High pressure warm liquid
K. Air handling unit
  L. Heating coil
  M. Fan
  N. Air from space
  O. Air to space
  P. Solar water heating coil
     (used only for solar assist)
  E. Expansion valve
  G. Check valve
S. Heat exchanger
   (used only for solar assist)

Solar Assisted Heating
A. Heat pump
  B. Compressor
  C. Condenser coil (not used
     with solar assist)
  D. Condenser fan
  E. Expansion valve
  F. Solenoid valve
  G. Typical check valve
  H. Diverter valve (cooling to
     heating cycles etc.)
  I. High pressure hot gas
  J. High pressure warm liquid
K. Air handling unit
  L. Heating coil
  M. Fan
  N. Air from space
  O. Air to space
  P. Solar water heating coil
  F. Expansion valve
  G. Check valve
S. Heater exchanger — for solar
   assist (refrig. gas to water)
  F. Solenoid valve
  E. Expansion valve
  V. Cold gas to heat exchanger
  W. Warm gas to compressor

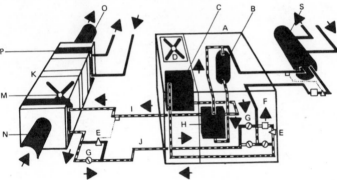

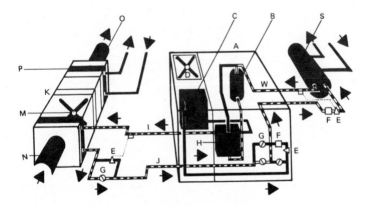

SOLAR ASSISTED HEAT PUMP

196

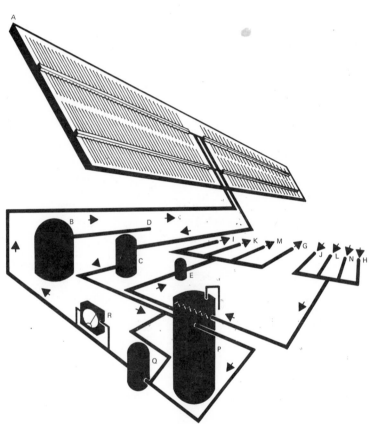

SOLAR COLLECTOR SYSTEM

A. Solar collector
B. Compression tank
C. Air separator
D. City water
E. Energy circulating pump
G. To domestic hot water system
H. From domestic hot water system

I. To perimeter finned
   tube radiation
J. From perimeter finned radiation
K. To hi-temp water absorption
   refrigeration system
L. From hi-temp water absorption
   refrigeration system

M. To cooling tower basin
N. From cooling tower basin
   heating system
P. Thermal storage tank
Q. Collector pump
R. Flow meter

Northeast elevation

South elevation

HOME SAVINGS & LOAN ASSOCIATION

LOCATION: Stoughton, Wisconsin.
LATITUDE: 43 degrees N.
REGION: Cool.
ARCHITECT: Don Reppen, Madison, Wisconsin.

SPACIAL SOLUTION

The Home Savings & Loan Office is Stoughton's first solar heated building. It was designed to encourage the clients to use solar energy and to present them with a real-life application. The architectural detailing follows a Norwegian style, commonly accepted in the town as a residential decoration. The fixed wood awnings were frequently used in Norwegian homes during the late 1800's. In addition to their aesthetic value, the awnings block direct summer light but let the winter rays through the double glazed windows. All the entrances have heat saving vestibules and are weather-stripped. The wood frame construction is insulated with 1 inch of styrofoam and fiberglass in the walls, and the roof is insulated with 12 inches of blown-in insulation.

MECHANICAL SOLUTION

The bank uses approximately 350 square feet of air-cooled, flat-plate collectors which are integrated with the roof. Air is circulated through the 22 panels by means of a small motor driven fan that automatically controls the early morning flow of air to the rooms. In the afternoon, air is drawn from the stone-filled storage tank in the basement. A conventional air duct distribution system is used, and an air to water heat exchanger preheats the domestic hot water in a separate electrically backed storage tank.

The solar energy system provides 70 percent of both the space and the domestic water heating needs. The make-up energy is supplied by a gas fired furnace.

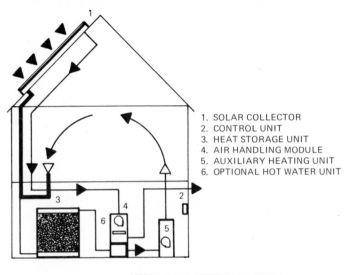

1. SOLAR COLLECTOR
2. CONTROL UNIT
3. HEAT STORAGE UNIT
4. AIR HANDLING MODULE
5. AUXILIARY HEATING UNIT
6. OPTIONAL HOT WATER UNIT

MECHANICAL SYSTEM SCHEMATIC

North elevation

East elevation

South elevation

199

Section 10. Temperate Region

The buildings presented in this section are the following:

Iris Images Inc. Film Processing Laboratory. Mill Valley, California.

San Francisco Environmental Center. San Francisco, California.

Cherry Creek Solar Office Building. Denver, Colorado.

Gump Glass Plant. Denver, Colorado.

Bross Utilities Service Corporation. Bloomfield, Connecticut.

Solar Office Building. Stamford, Connecticut.

Medical Office Building. Wichita, Kansas.

Solar Office Building. Mead, Nebraska.

Concord National Bank. Concord, New Hampshire.

Friendship Federal Bank. Greensburg, Pennsylvania.

Friendship Federal Bank. Ingomar, Pennsylvania.

Custom Leather Boutique. White River Junction, Vermont.

Eastern Liberty Savings Bank. Washington, D.C.

Hogate's Restaurant. Washington, D.C.

IRIS IMAGES INC. FILM PROCESSING LAB.

LOCATION: Mill Valley, California.
LATITUDE: 37.9 degrees N.
REGION: Temperate.
DESIGNER AND BUILDER: Iris Images, Mill Valley, California.

MECHANICAL SOLUTION

This project is concerned with the integration of a solar water heating system to a film processing operation. The laboratory's four film processors are estimated to require 12 gallons per minute of 100-degree-F water, 60 percent of which is provided by the 640-square-foot, flat-plate, water-cooled collectors mounted on tilted decks on the flat roof. Water is only circulated when the collector temperature is higher than that of the 360-gallon capacity storage tank. A 100-gallon gas fired water heater is used as a back-up.

Interior

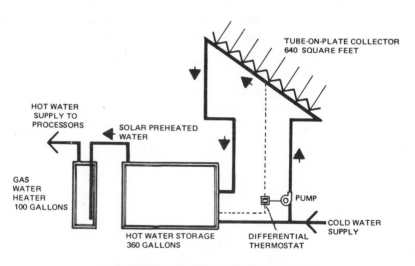

SOLAR ASSISTED INDUSTRIAL WATER HEATING SYSTEM

Front elevation

203

SAN FRANCISCO ENVIRONMENTAL CENTER

LOCATION: San Francisco, California.
LATITUDE: 37.5 degrees N.
REGION: Temperate.
ARCHITECT: Storek & Storek, San Francisco, California.
SOLAR CONSULTANT: Berkeley Solar Group, Berkeley, California.

SPACIAL AND MECHANICAL SOLUTION

The environmental center is a two-face construction project involving the rehabilitation of a 1916 power plant and its new addition. The combination provides 100,000 square feet of prime location offices, retail shops, restaurants and tenant storage.

Phase I of the project, which included the new addition to the power plant, provided 32,000 square feet of office space. Both passive and active solar energy systems are used. The structure is of steel and reinforced concrete, with 18,000 square feet of liquid-cooled, flat-plate collectors integrated into the exterior skin. A rooftop garden shades the building during the summer and doubles as a gathering place. Natural ventilation provides most of the cooling. For mechanical cooling, done at low demand hours, the cold water storage is used.

Phase II of the project included the power plant rehabilitation, which provided 46,000 square feet of offices and made use of the passive thermal qualities inherent in the building's construction (massive brick and terra cotta walls and a thick concrete slab).

Both structures make use of zoned lighting, which is brighter in the work areas and darker elsewhere. Zoned air systems with different gas heat compressor units serve the four climate areas: south, north, top and interior.

Perhaps the most innovative building components used are the sun tracking mirrors, or helistats, used to bounce sunlight into the interior spaces. The windows supply 23 percent of the annual energy delivered to the building for space heating; the solar collectors provide 57 percent and the gas fired furnace supplies the remaining 20 percent.

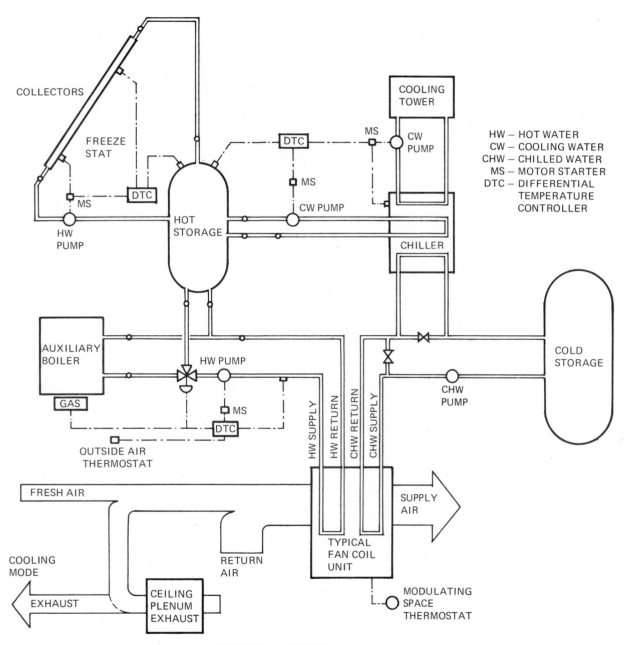

COLLECTORS

FREEZE
STAT

DTC

MS

COOLING
TOWER

CW
PUMP

HW — HOT WATER
CW — COOLING WATER
CHW — CHILLED WATER
MS — MOTOR STARTER
DTC — DIFFERENTIAL
TEMPERATURE
CONTROLLER

MS

DTC

MS

HOT
STORAGE

CW PUMP

HW
PUMP

CHILLER

AUXILIARY
BOILER

HW PUMP

COLD
STORAGE

GAS

MS

CHW
PUMP

DTC

OUTSIDE AIR
THERMOSTAT

HW SUPPLY

HW RETURN

CHW RETURN

CHW SUPPLY

FRESH AIR

SUPPLY
AIR

COOLING
MODE

RETURN
AIR

TYPICAL
FAN COIL
UNIT

EXHAUST

CEILING
PLENUM
EXHAUST

MODULATING
SPACE
THERMOSTAT

MECHANICAL SYSTEM DIAGRAM

205

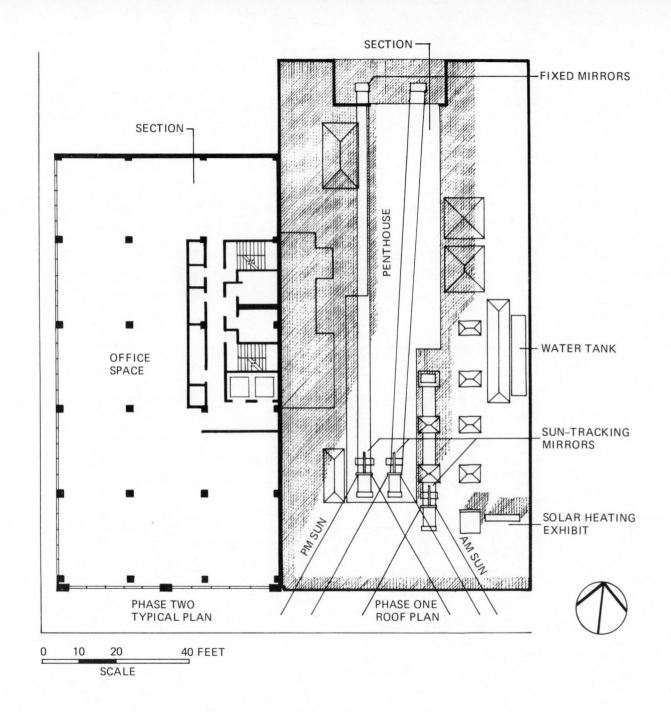

SECTION

FIXED MIRRORS

SECTION

PENTHOUSE

WATER TANK

SUN-TRACKING
MIRRORS

OFFICE
SPACE

PM SUN

AM SUN

SOLAR HEATING
EXHIBIT

PHASE TWO
TYPICAL PLAN

PHASE ONE
ROOF PLAN

0   10   20   40 FEET

SCALE

206

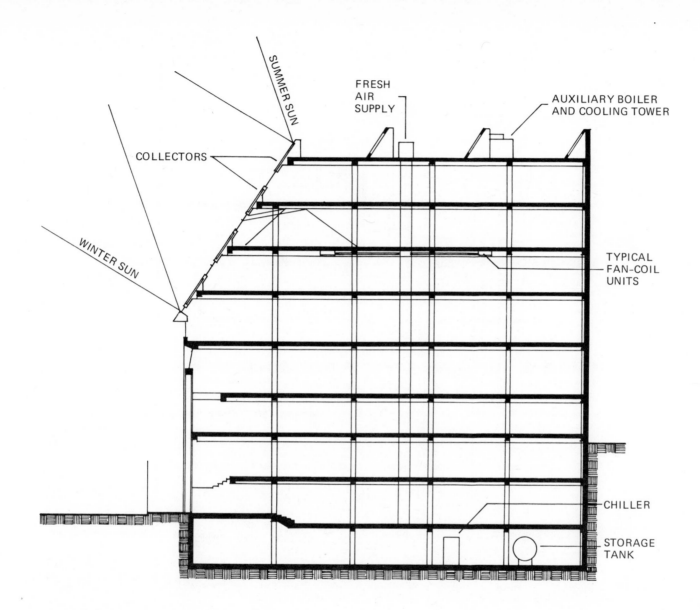

SUMMER SUN

FRESH
AIR
SUPPLY

AUXILIARY BOILER
AND COOLING TOWER

COLLECTORS

WINTER SUN

TYPICAL
FAN-COIL
UNITS

CHILLER

STORAGE
TANK

BUILDING SECTION-PHASE TWO

0   10   20   40 FEET

SCALE

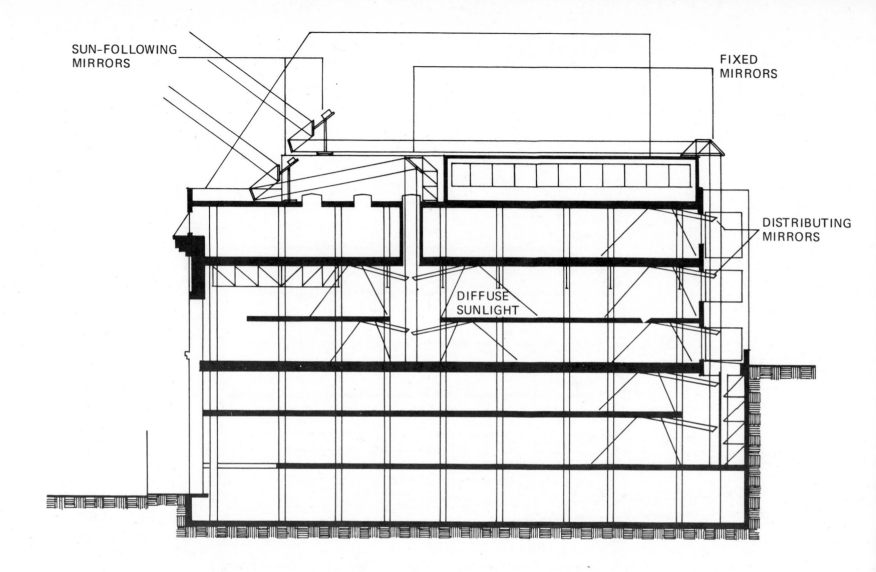

SUN-FOLLOWING
MIRRORS

FIXED
MIRRORS

DISTRIBUTING
MIRRORS

DIFFUSE
SUNLIGHT

BUILDING SECTION–PHASE ONE

0   10   20   40 FEET

SCALE

208

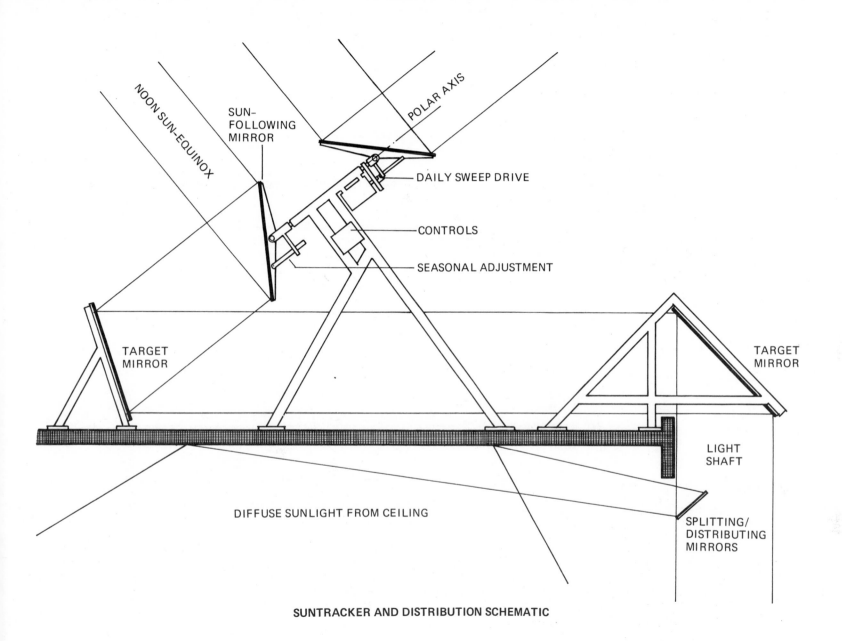

NOON SUN-EQUINOX

SUN-FOLLOWING MIRROR

POLAR AXIS

DAILY SWEEP DRIVE

CONTROLS

SEASONAL ADJUSTMENT

TARGET MIRROR

TARGET MIRROR

LIGHT SHAFT

DIFFUSE SUNLIGHT FROM CEILING

SPLITTING/ DISTRIBUTING MIRRORS

SUNTRACKER AND DISTRIBUTION SCHEMATIC

209

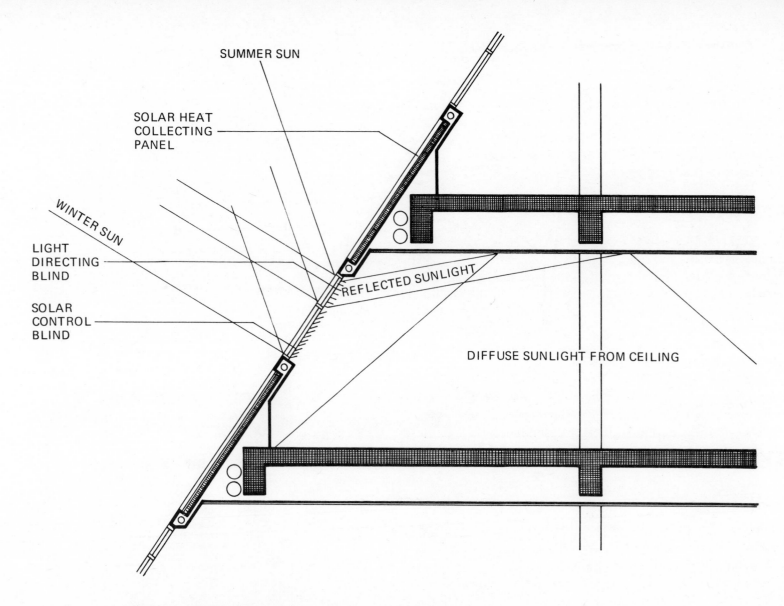

SUMMER SUN

SOLAR HEAT
COLLECTING
PANEL

WINTER SUN

LIGHT
DIRECTING
BLIND

SOLAR
CONTROL
BLIND

REFLECTED SUNLIGHT

DIFFUSE SUNLIGHT FROM CEILING

DETAIL–SCHEMATIC SECTION THROUGH SOUTH FACE

0   2   4   8   16 FEET

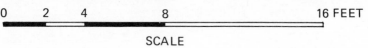

SCALE

210

CHERRY CREEK SOLAR OFFICE BUILDING

LOCATION: Denver, Colorado.
LATITUDE: 40 degrees N.
REGION: Temperate.
ARCHITECT: Richard Crowther, Denver, Colorado.
MECHANICAL AND ELECTRICAL ENGINEER: Walto
    Associates, Denver, Colorado.

SPACIAL SOLUTION

This project reflects a sensitive blend of energy conscious design, energy management and solar technology.

Primary consideration was given to the building's orientation and shape, the form thickness of the exterior walls, the location of entries and natural ventilation. The majority of the windows and entries are located on the south elevation to obtain maximum exposure to the winter sun and for protection against the winter winds. Berms funnel prevailing winds away from the exterior surfaces.

The first level of the two-story building is partly below grade to reduce heat losses. The use of a wooden foundation permits the insulation to be continuous from the foundation plate to the top of the roof parapet.

MECHANICAL SOLUTION

The building uses 136 square feet of flat-plate, air-cooled collectors facing directly south and tilted to 45 degrees from the horizontal. White marble chips on the flat roof and white stucco panels reflect light into the collector. There is a clerestory window above the collector, allowing direct sunlight to enter the building. During the summer months, the reflecting panel shades the clerestory.

The solar system provides space heating. Fans are used to recycle warm air from the ceiling to the floor. A heat pump is used

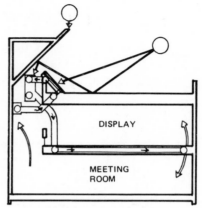

HEATING WITH THE COLLECTOR

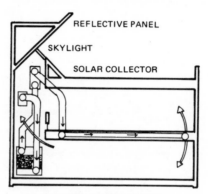

REFLECTIVE PANEL

SKYLIGHT

SOLAR COLLECTOR

HEATING FROM STORAGE

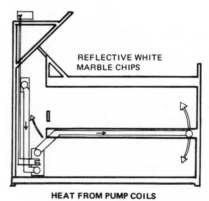

REFLECTIVE WHITE
MARBLE CHIPS

HEAT FROM PUMP COILS

in parallel with—but independent of—the solar system. A storage bin with 3 tons of gravel stores the thermal energy.

The solar system provides 20 percent of the heating needs. However, the building's energy needs have been reduced by 80 percent due to efficient design. The required 20 percent energy make-up is supplied by an electric heating system.

Southwest elevation

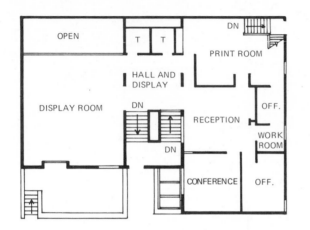

FLOOR PLANS

Central passageway

GUMP GLASS PLANT

LOCATION: Denver, Colorado.
LATITUDE: 40 degrees N.
REGION: Temperate.
ARCHITECT: Intergroup, Denver, Colorado.

MECHANICAL SOLUTION

Of the total 41,000 square feet of floor area available in this plant, 7,200 square feet are conditioned by the solar system. The offices and showroom are located in this area.

The solar system is composed of 1,600 square feet of air-cooled, flat-plate collectors mounted on a flat roof in four rows and tilted to 50 degrees from the horizontal. The storage bin contains 60 tons of gravel and is adjacent to the office area. A

Roof mounted collectors

blower circulates the heated air from the storage (or from the collector, when the storage doesn't have enough thermal energy) to the five separately controlled office zones. The solar system provides over 85 percent of the heating needs for the space it conditions. Domestic hot water, space cooling and back-up space heating is powered by gas.

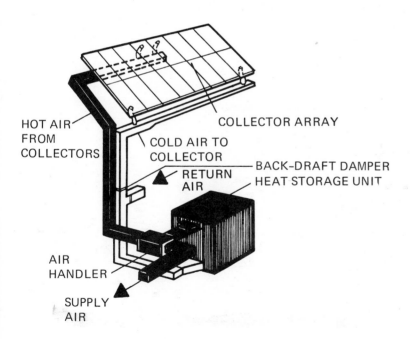

HOT AIR FROM COLLECTORS

COLLECTOR ARRAY

COLD AIR TO COLLECTOR

BACK-DRAFT DAMPER
HEAT STORAGE UNIT

RETURN AIR

AIR HANDLER

SUPPLY AIR

SOLAR SYSTEM SCHEMATIC

213

BROSS UTILITIES SERVICE CORPORATION

LOCATION: Bloomfield, Connecticut.
LATITUDE: 41.5 degrees N.
REGION: Temperate.
DESIGNER AND BUILDER: Bross Utilities Service Corporation, Bloomfield, Connecticut.
MECHANICAL ENGINEER: Koton Engineering, Bloomfield, Connecticut.

MECHANICAL SOLUTION

This retrofit, two-story, 12,800-square-foot office building uses the solar system as its main energy providing source. The solar collection system consists of 2,430 square feet of liquid-cooled, flat-plate collectors, supported by a steel frame which is mounted on a flat roof and tilted 52 degrees from the horizontal. The last four collector arrays are elevated 5 feet to allow the required collector area to fit in the limited roof area. The heat removal agent is circulated from the collector to a 5,000-gallon storage tank by a 3-horsepower centrifugal pump. The system

West elevation

preheats the domestic hot water and powers the 3-ton absorption chiller and the 25-ton absorption chiller that are used for cooling. Sixty percent of the total annual energy needs are supplied by the solar system; the remaining energy demand is supplied by an existing oil and electric furnace.

Southwest elevation

SOLAR OFFICE BUILDING

LOCATION: Stamford, Connecticut.
LATITUDE: 41.1 degrees N.
REGION: Temperate.
OWNER AND BUILDER: Lutz Sotire Partnership, Stamford,
    Connecticut.

Southeast elevation

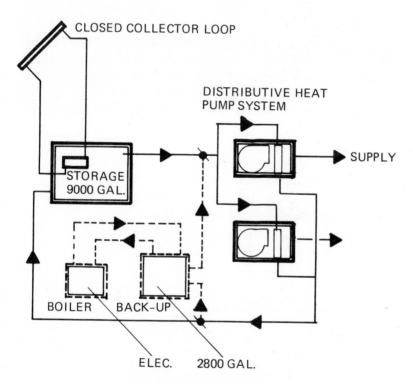

SOLAR SYSTEM SCHEMATIC

MECHANICAL SOLUTION

This project entailed the retrofitting of a seven-year-old office building to utilize solar space heating. The three-story, 25,000-square-foot building has 3,174 square feet of liquid-cooled, flat-plate collectors, mounted on steel frames in seven rows and tilted 55 degrees from the horizontal. Directly south of each collector bank and facing north, there is a reflector sheet tilted to 10 degrees from the horizontal. It is estimated that this increases energy collection by 42 percent. The 9,000-gallon storage tank was placed at the rear of the building, beneath the old parking lot. The existing closed loop, multiple zone heat pump was integrated into the solar system, which provides 100 percent of the energy needed for space heating.

215

MEDICAL OFFICE BUILDING

LOCATION: Wichita, Kansas.

LATITUDE: 37 degrees N.

REGION: Temperate.

ARCHITECT: Gossen Livingstone Associates, Wichita, Kansas.

SPACIAL SOLUTION

This one-story office building has 4,764 feet of floor area, including a basement with a 1,380-square foot area. The main entrance has a double glazed door and a vestibule that is recessed into an interior court. All the windows are double glazed and comprise less than 7 percent of the total exterior area.

MECHANICAL SOLUTION

The solar system was designed exclusively for space heating. The 1,160 square feet of air-cooled, flat-plate collectors were installed in place of the south wall's outside cover. The need for additional protective wall surface material was eliminated by attaching the absorber plate directly to the plywood sheading; this permits the wall insulation to double as collector insulation. The storage contains a change of phase and eutectic salts stored in trays and insulated by 12 inches of fiberglass. The salts have five times more heat storage capacity in the same volume than does water. Hot air from the collector is blown over the tray containers and the salts absorb the heat. Cool air is circulated through the stored tray to warm the air and direct it back to the rooms for heating. The combination of passive features and the active solar system provides 90 percent to 100 percent of the heating requirements. The make-up energy, if any, is supplied by an electric furnace.

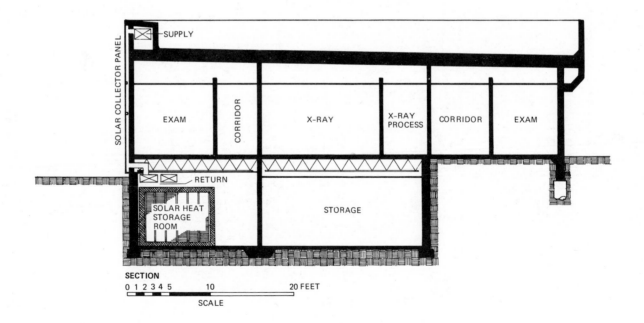

SECTION

0 1 2 3 4 5   10   20 FEET

SCALE

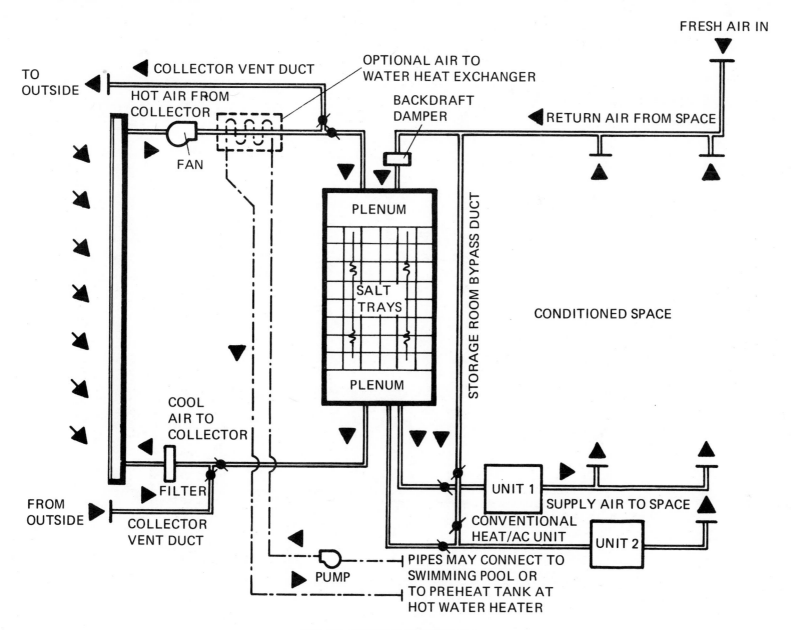

SOLAR SYSTEM SCHEMATIC

TO OUTSIDE

COLLECTOR VENT DUCT

HOT AIR FROM COLLECTOR

OPTIONAL AIR TO WATER HEAT EXCHANGER

FRESH AIR IN

BACKDRAFT DAMPER

RETURN AIR FROM SPACE

FAN

PLENUM

SALT TRAYS

PLENUM

STORAGE ROOM BYPASS DUCT

CONDITIONED SPACE

COOL AIR TO COLLECTOR

FILTER

FROM OUTSIDE

COLLECTOR VENT DUCT

PUMP

UNIT 1

SUPPLY AIR TO SPACE

CONVENTIONAL HEAT/AC UNIT

UNIT 2

PIPES MAY CONNECT TO SWIMMING POOL OR TO PREHEAT TANK AT HOT WATER HEATER

217

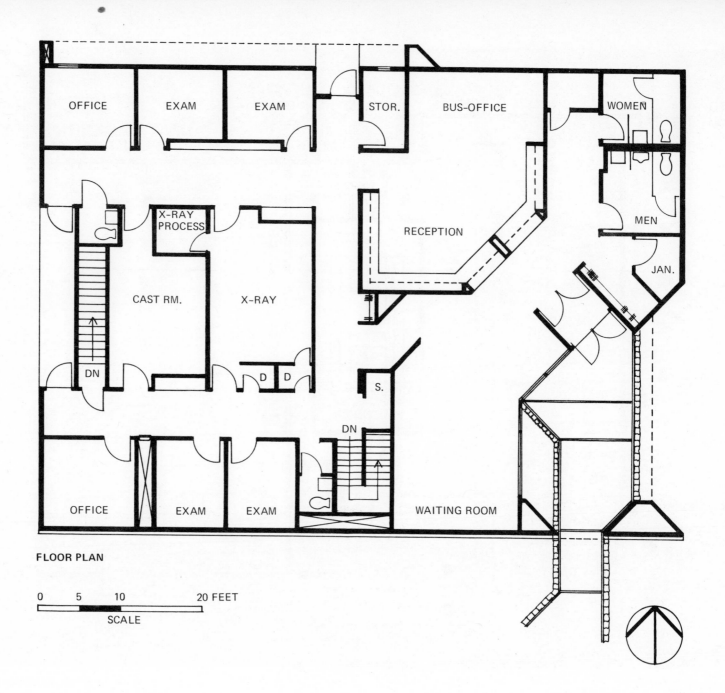

FLOOR PLAN

```
0    5   10           20 FEET
SCALE
```

Southeast elevation

Photo by James Lemkin

Southwest elevation

Photo by James Lemkin

Interior of mechanical room showing solar storage room.

Photo by James Lemkin

SOLAR OFFICE BUILDING

LOCATION: Mead, Nebraska.
LATITUDE: 41 degrees N.
REGION: Temperate.
ARCHITECT AND SOLAR ENGINEER: Hansen, Lynd and
     Meyer, Iowa City, Iowa.

SPACIAL SOLUTION

This two-story, 1,764-square-foot office building was designed
for a manufacturer of solar equipment who wanted to combine
the need for new office space with a demonstration of the company's product. The building incorporates many active and passive energy conserving and producing features. Both the walls
and the ceilings are heavily insulated with a *U* value of 0.05. The
main entrance consists of a vestibule that reduces infiltration
and heat loss. Windows are double glazed and horizontal in
shape. This effectively reduces the over-all window area and the
heat losses.

MECHANICAL SOLUTION

The solar system was designed to exclusively provide space heating. The 800 square feet of air-cooled, flat-plate collectors are an
integral part of the vertical south wall. Air is circulated through
a galvanized iron absorber plate by a 1-horsepower blower. A 5-
ton storage system, located on the second floor and using glau-
ber salts, stores the heat absorbed by the collector. The storage
bin is insulated with 6 inches of dylite perma foam. Air is circulated to the rooms by a separate 3/8-horsepower blower.

   The combination of passive and active systems provides 90
percent to 100 percent of the space heating requirements. The
make-up energy is supplied by an electric furnace.

CONCORD NATIONAL BANK

LOCATION: Concord, New Hampshire.
LATITUDE: 43 degrees N.
REGION: Temperate.
ARCHITECT: Kenneth Parry Associates, Quincy,
     Massachusetts.
MECHANICAL ENGINEER: C. A. Crowley Engineering,
     Brockton, Massachusetts.

SPACIAL SOLUTION

This one-story, 1,900-square-foot bank is of masonry construction, with walls made of 4-inch brick, 8-inch block, 1.5-inch
rigid urethene foam and 5/8-inch gypsum board, for a total resistance of $R=11$. The built-up roof is made of 5/8-inch plywood sheading, 1.5-inch rigid urethane foam insulation, and 6
inches of fiberglass insulation placed in between the rafters, for
a total resistance of $R=28$. All windows are single glazed and
are protected from the summer sun by deep overhangs.

MECHANICAL SOLUTION

This solar system is considered to be the first to provide both
heating and cooling in the New England area.

   The 575 square feet of liquid-cooled, flat-plate collectors are
mounted as an integral part in the south-facing roof and are
sloped to 45 degrees from the horizontal. Four heat dump
panels are included in the array to prevent overheating of the
other panels during hot summer days. The heat removing agent,
a combination of ethylene glycol and 60 percent water, is circulated from the collectors to the heat exchanger in the storage
tank by a centrifugal pump.

The cylindrical steel storage tank, with its 1,500-gallon capacity, is insulated with 4 inches of spray foam and is buried at the building's south end. Rooms are heated through fan-coil units in the forced air system, and are cooled by a solar powered 5-ton absorption chiller.

Solar energy supplies 70 percent of the total heating and cooling requirements. The make-up energy is provided by a heat pump and electric coils in the air ducts.

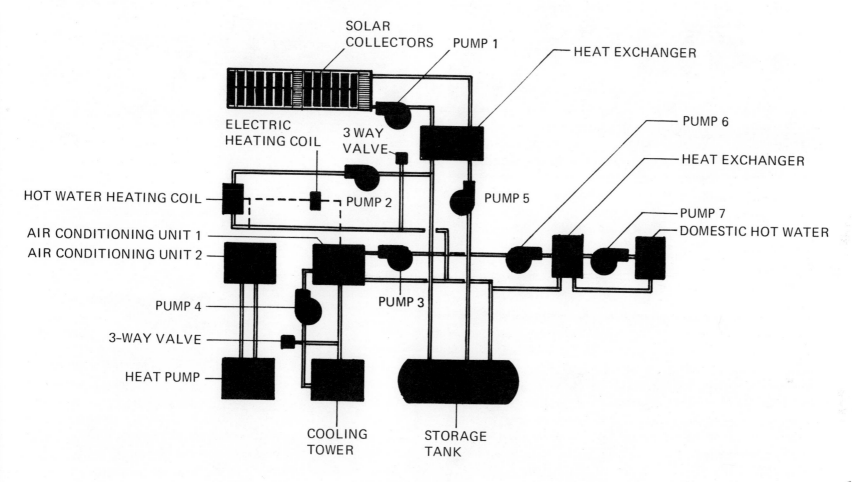

FLOW DIAGRAM

Southwest elevation

North elevation

South elevation

224

Entrance detail

FRIENDSHIP FEDERAL BANK

LOCATION: Greensburg, Pennsylvania.
LATITUDE: 41 degrees N.
REGION: Temperate.
ARCHITECT AND SOLAR DESIGNER: Burt, Hill & Associates,
    Butler, Pennsylvania.

SPACIAL SOLUTION

This 2,000-square-foot facility utilizes several energy conscious features, in addition to the solar system, to reduce the over-all energy consumption. The walls and ceiling are substantially insulated. The exterior walls are partially bermed to reduce the effect of cold winter winds. All windows are double glazed and are deeply recessed to avoid penetration of direct summer sunlight. The main entrance is from the east and uses a double door vestibule.

MECHANICAL SOLUTION

The bank obtains space and domestic water heating energy from 26 Pittsburg Plate Glass (PPG), flat-plate, liquid-cooled panels, mounted on two wooden frame arrays and tilted to 45 degrees from the horizontal. The frames are made with 2-inch by 4-inch trusses and sealed with aluminum flashing. There is no sheathing on the trusses; the side facing north has 1-inch by 6-inch wood louvers and the ends are covered with plywood. The heat removal agent is stored in an insulated, 1,000-gallon capacity concrete tank and it is circulated through a coil in the return air duct of the gas fired furnace. If the solar system is not able to heat the air to the required temperature, the furnace automatically switches on to provide the needed boost. The storage tank is also used to preheat the domestic hot water. The bank utilizes the solar system to provide over 60 percent of the space and water heating needs.

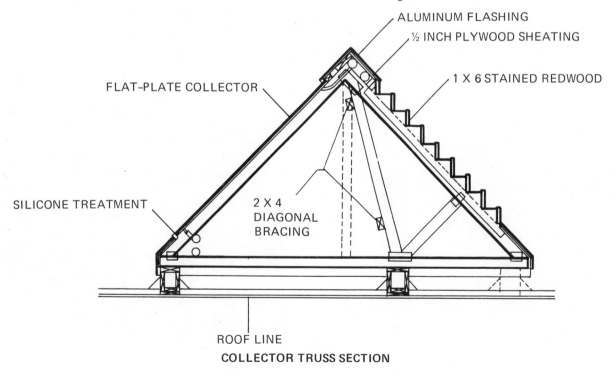

ALUMINUM FLASHING
½ INCH PLYWOOD SHEATING
1 X 6 STAINED REDWOOD
FLAT-PLATE COLLECTOR
SILICONE TREATMENT
2 X 4 DIAGONAL BRACING
ROOF LINE

COLLECTOR TRUSS SECTION

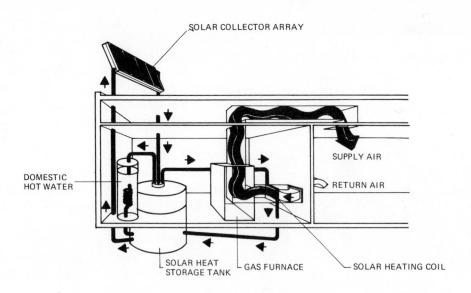

SOLAR COLLECTOR ARRAY

DOMESTIC
HOT WATER

SUPPLY AIR

RETURN AIR

SOLAR HEAT
STORAGE TANK

GAS FURNACE

SOLAR HEATING COIL

SOLAR SYSTEM SCHEMATIC

Southeast elevation

Interior

FRIENDSHIP FEDERAL BANK

LOCATION: Ingomar, Pennsylvania.
LATITUDE: 41 degrees N.
REGION: Temperate.
ARCHITECT AND SOLAR DESIGNER: Burt, Hill & Associates, Butler, Pennsylvania.

SPACIAL SOLUTION

The gas company denied the hook-up permit for this project, and a look at prospective costs of an all-electric building made the project monetarily unfeasible, leaving solar energy as the only alternative. Encouraged by the more than favorable results produced by a branch office, which uses solar energy, Friendship Federal decided to make their Ingomar branch their second solar heated facility.

The two-story, 4,500-square-foot building provides a lobby, work areas, offices and space for supportive banking functions, as well as 2,000 square feet of office rental space on the second floor. Its scale and appearance is suitable for the suburban community in which it is located. Some of the energy saving features include the use of the natural site slope for the insulating value of the berm. Steel reinforced construction was used to add to the thermal mass, and a skylight, designed to provide the greatest amount of natural light and heat gain, was installed in the south facade.

MECHANICAL SOLUTION

The solar energy system provides space heating for the first floor only and solar water heating for both floors. The system can best be described as a solar assisted heat pump. It provides 60 percent of the space heating and 80 percent of the water heating, with 540 square feet of PPG flat-plate collectors and a 1,000-gallon capacity storage tank.

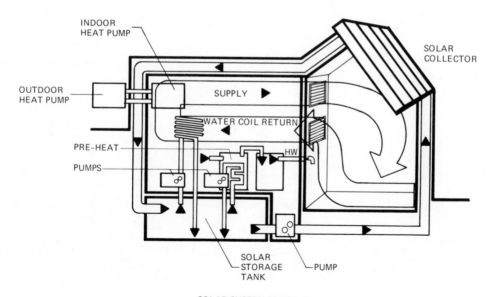

SOLAR SYSTEM SCHEMATIC

227

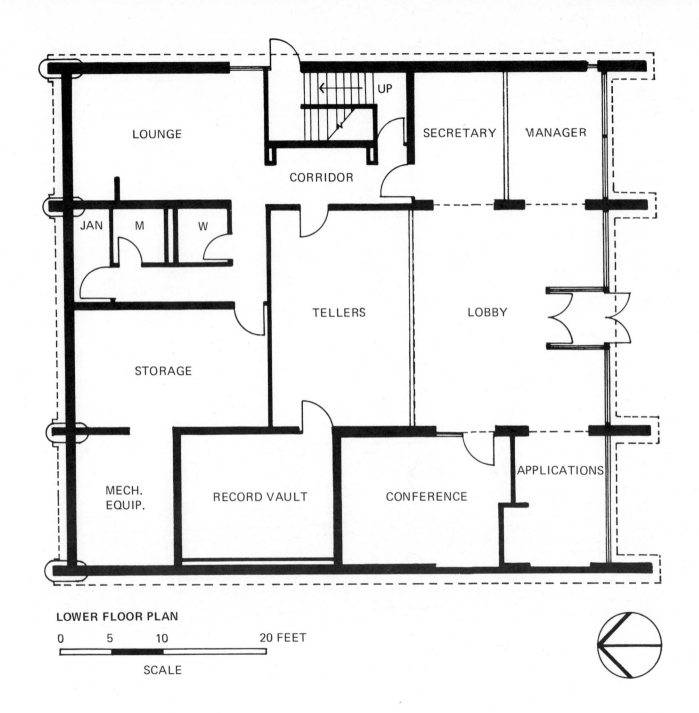

LOWER FLOOR PLAN

0   5   10   20 FEET

SCALE

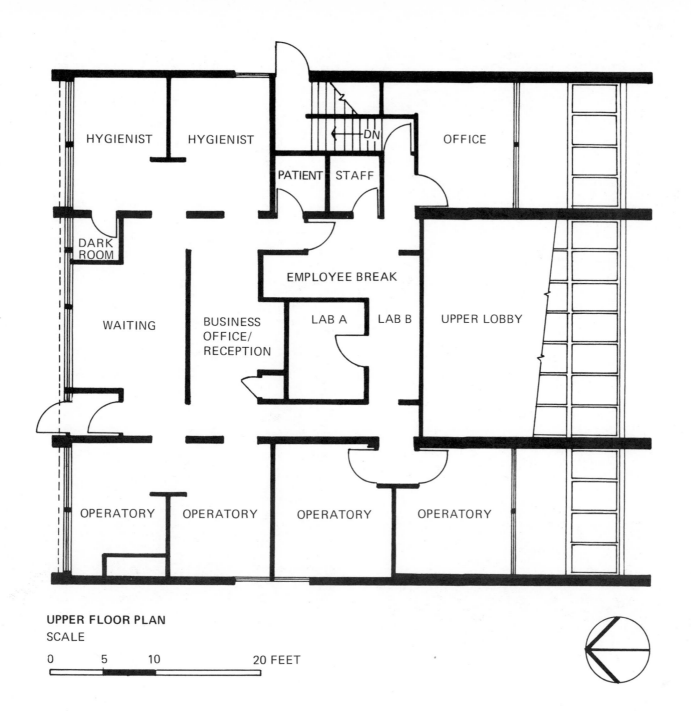

HYGIENIST   HYGIENIST   OFFICE

                                    PATIENT   STAFF

DARK
ROOM

                                    EMPLOYEE BREAK

WAITING   BUSINESS   LAB A   LAB B   UPPER LOBBY
              OFFICE/
              RECEPTION

OPERATORY   OPERATORY   OPERATORY   OPERATORY

UPPER FLOOR PLAN
SCALE

0   5   10   20 FEET

Southwest elevation

Interior skylight

Northeast elevation

230

CUSTOM LEATHER BOUTIQUE

LOCATION: White River Junction, Vermont.
LATITUDE: 43.5 degrees N.
REGION: Temperate.
SOLAR ENGINEER: Sol-R-Tech, Denver, Colorado.

SPACIAL AND MECHANICAL SOLUTION

This one-story, wooden frame, concrete slab, retail store has 2,100 square feet of usable floor area. Its walls are insulated with 4 inches of fiberglass and the ceiling has 6 inches of the fiberglass insulation. The main entrance is from the north, where there are two double glazed show windows. There are no windows in the east or south elevations and a door connects the store with the tennis courts through the south wall.

A new building, constructed adjacent to the west end of the boutique, casts an afternoon shadow on a portion of the 910 square feet of collectors. However, this only reduces the system's efficiency by a few percentage points. The collectors are flat-plate and water-cooled and are integrated with the roof's structural rafters at a tilt of 45 degrees from the horizontal. The heat removal agent is water and corrosion inhibitors, with no anti-freeze needed since the system is equipped with an automatic draining valve. The storage tank has a 3,200-gallon capacity and it operates in combination with three York water to air heat pumps and fan-coil units in the forced air system. The solar energy system, with the heat pumps, provides 80 percent of the space heating needs; the remaining energy requirements are supplied by electric heating elements in the ducts.

Southeast elevation

EASTERN LIBERTY SAVINGS BANK

LOCATION: Washington, D.C.
LATITUDE: 39 degrees N.
REGION: Temperate.
ARCHITECT: Mills, Petticord Partnership, Washington, D.C.
ENGINEER: Syska & Henessey, New York, New York.

MECHANICAL SOLUTION

The 2,000 square feet of liquid-cooled, flat-plate, PPG collectors are integrated into the building's second mansard roof, which is tilted to 45 degrees from the horizontal. The heat removal agent flows from the collectors to the heat exchangers in the storage tank located in the mechanical penthouse. Electrical heaters in the tank provide back-up energy when the solar system is not able to meet the heating requirements. A water to air pump placed under each window, and assisted by the solar heated water, provides perimeter heating. The interior zone is heated with a conventional, electric, forced air system. The domestic hot water is preheated by the storage tank.

Southwest elevation

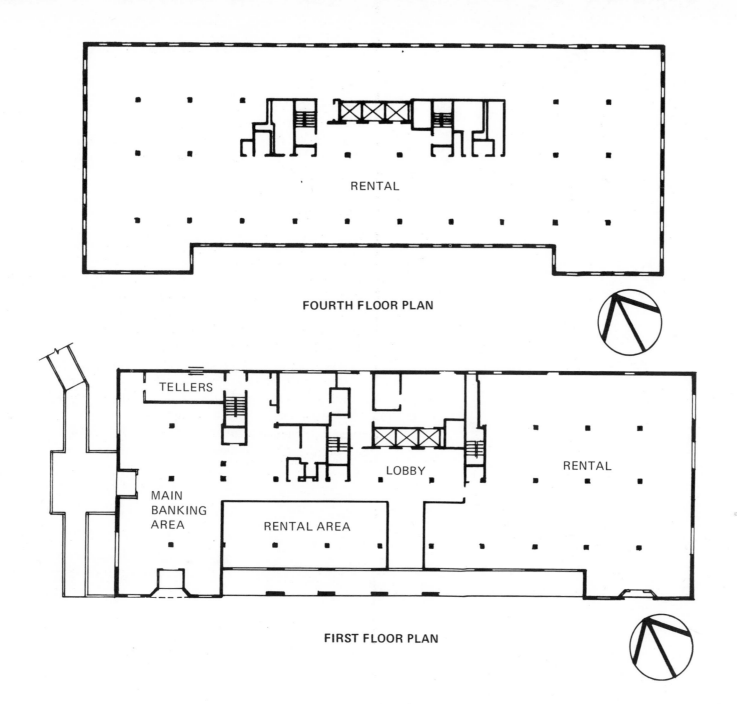

FOURTH FLOOR PLAN

RENTAL

TELLERS

LOBBY

RENTAL

MAIN
BANKING
AREA

RENTAL AREA

FIRST FLOOR PLAN

233

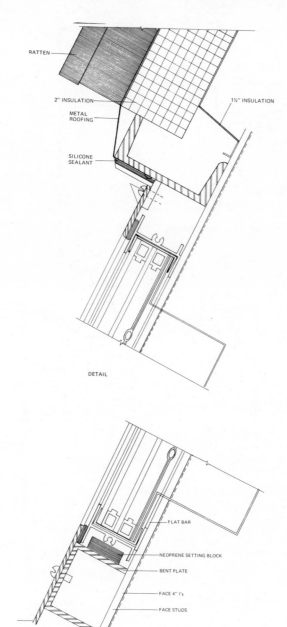

RATTEN

2" INSULATION

METAL ROOFING

SILICONE SEALANT

1½" INSULATION

DETAIL

FLAT BAR

NEOPRENE SETTING BLOCK

BENT PLATE

FACE 4" I's

FACE STUDS

COLLECTOR DETAIL

HOGATE'S RESTAURANT

LOCATION: Washington, D.C.
LATITUDE: 39 degrees N.
REGION: Temperate.
SOLAR ENGINEER: Muller Associates, Baltimore, Maryland.
GENERAL CONTRACTOR AND OWNER: Marriott Corporation, Washington, D.C.

MECHANICAL SOLUTION

The 5,700 square feet of flat-plate, liquid-cooled collectors were placed on the roof of the existing restaurant. The system was designed to supplement the gas fired water heater, which supplies 10,000 gallons of hot water every day.

The collector arrays were designed around the existing roof configuration, which did not allow for the optimal tilt angle to be used. The heat removal agent is a combination of ethylene glycol and water, and it circulates from the collectors to the heat exchangers located in the attic. Water is circulated from the two 5,000-gallon storage tanks located in the basement, to the heat exchangers and then to the gas fired water heater, which increases the temperature when needed. The solar system provides 55 percent of the domestic water heating requirements.

Roof mounted collectors

South elevation

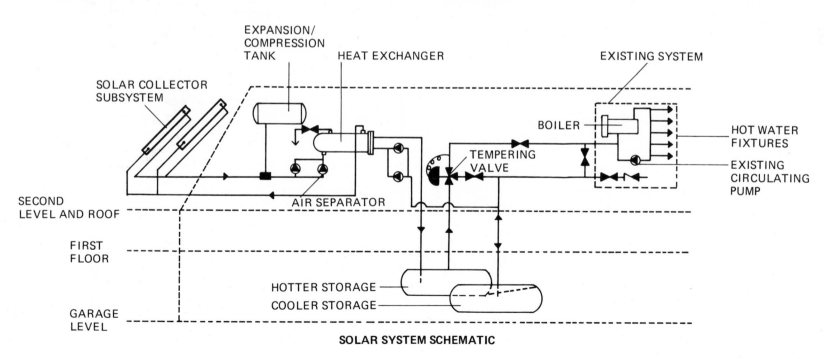

SOLAR COLLECTOR SUBSYSTEM

EXPANSION/ COMPRESSION TANK

HEAT EXCHANGER

EXISTING SYSTEM

BOILER

HOT WATER FIXTURES

TEMPERING VALVE

EXISTING CIRCULATING PUMP

AIR SEPARATOR

SECOND LEVEL AND ROOF

FIRST FLOOR

HOTTER STORAGE

COOLER STORAGE

GARAGE LEVEL

SOLAR SYSTEM SCHEMATIC

Section 11. Hot Arid Region

The buildings presented in this section are the following:

Gila River Ranch Irrigation Plant. Gila Bend, Arizona.

Bridges and Paxton Office Building. Albuquerque, New Mexico.

United Southwest National Bank. Sante Fe, New Mexico.

GILA RIVER RANCH IRRIGATION PLANT

LOCATION: Gila Bend, Arizona.
LATITUDE: 32 degrees N.
REGION: Hot Arid.
SYSTEM DEVELOPMENT: Battelle Memorial Institute,
    Columbus, Ohio.
OWNER: Northwest Life Insurance Company, Madison,
    Wisconsin.

MECHANICAL SOLUTION

Although this project does not currently use the harnessed solar energy for space conditioning, plans have been developed for alternate uses of the collected energy. Some of the projected off-season applications include the use of solar energy for the drying of grain and crops, and space heating for greenhouses, livestock shelters and farm houses. However, the primary function of the 5,500 square feet of sun-tracking, concentrating collectors is to power the irrigation system, which serves 25,000 acres of farm land. The nine collector arrays are arranged on the north-south axis and track the sun from east to west. The heated water, which reaches temperatures of 300 degrees F, drives a power package that heats freon into a gaseous state. This freon gas is expanded through a turbine which drives the irrigation pump. The pump lifts the water from the sump basin some 14 feet to the irrigation canal. The system has a maximum delivery capacity of 10,000 gallons per minute. During high winds and at night, the collectors are tilted downward to prevent damage to the reflector panels.

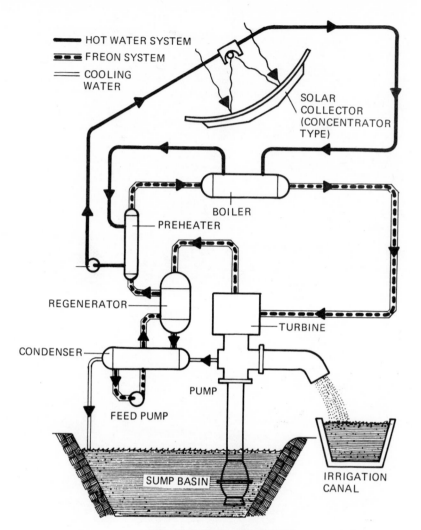

SOLAR SYSTEM SCHEMATIC

South view of collectors

Collectors in closing position

Collectors in open position

BRIDGES AND PAXTON SOLAR OFFICE BUILDING

LOCATION: Albuquerque, New Mexico.
LATITUDE: 35 degrees N.
REGION: Hot Arid.
SOLAR AND MECHANICAL ENGINEER: Bridges and Paxton
   Engineers, Albuquerque, New Mexico.
ARCHITECT: Stanley & Wright, Albuquerque, New Mexico.

MECHANICAL SOLUTION

In 1956, this building utilized solar energy to heat 4,300 square feet of office space and a drafting area. However, from 1962 to 1974, this heating system was non operational. During this time, an addition increased the building's floor area to 8,300 square feet, and a boiler and furnace were installed to provide space heating. In 1974, the solar energy system was reactivated, using improved components and automated operational controls to improve efficiency.

The revised mechanical system has 750 square feet of liquid-cooled, flat-plate collectors tilted to 60 degrees from the horizontal. An ethylene glycol and water combination is circulated from the collector to the heat exchanger in the 6,000-gallon storage tank. Five water to air heat pumps are used to heat five separately controlled zones. The solar energy system provides 100 percent of the space heating requirements of the 5,000 square foot office building. No back-up system is needed.

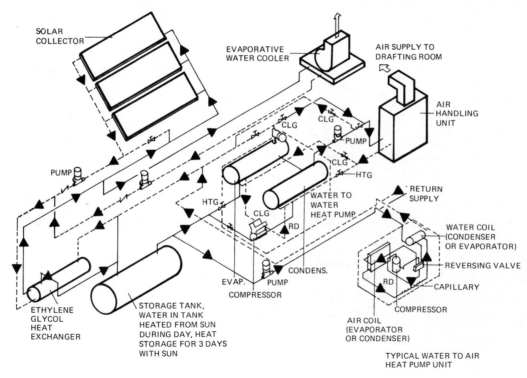

SOLAR HEATING SCHEMATIC

Side Elevation

Storage Tanks

UNITED SOUTHWEST NATIONAL BANK

LOCATION: Sante Fe, New Mexico.
LATITUDE: 36 degrees N.
REGION: Hot Arid.
ARCHITECT: Bernabe Romero, Sante Fe, New Mexico.
SOLAR ENGINEERING: Herma G. Barkman & Associates,
    Sante Fe, New Mexico.

MECHANICAL SOLUTION

This two-story, 5,000-square-foot bank follows the traditional adobe architectural style of the southwest. Its solar energy system consists of 980 square feet of liquid-cooled, flat-plate collectors mounted in five rows, supported by wooden frames and tilted to 55 degrees from the horizontal. The heat removal medium is a combination of water and ethylene glycol, which flows from the collector to the heat exchanger outside the 7,000-gallon tank. Water is circulated from the tank to fan-coil units in a forced air system. The bank is divided into five separately controlled heating zones to maximize energy conservation. The solar system provides 60 percent of the space heating needs, and a boiler supplies the make-up energy.

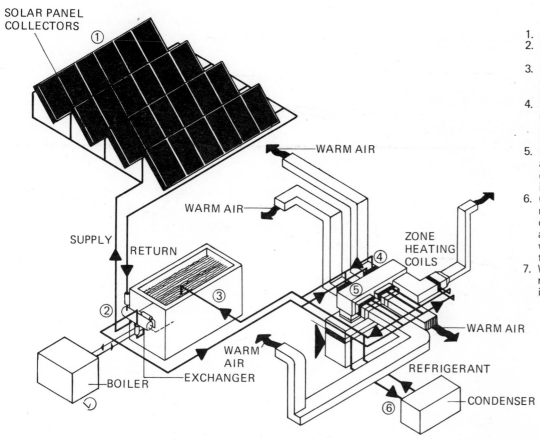

SOLAR PANEL COLLECTORS

WARM AIR

WARM AIR

SUPPLY

RETURN

ZONE HEATING COILS

WARM AIR

WARM AIR

BOILER

EXCHANGER

WARM AIR

REFRIGERANT

CONDENSER

SOLAR SYSTEM SCHEMATIC

1. The solar collectors absorb heat from the sun's rays.
2. Circulating anti-freeze solution picks up heat from the collector and takes it to the heat exchanger.
3. The heat exchanger transfers the heat from the antifreeze solution to circulating water which is stored in the insulated storage tank.
4. The hot water from the heat storage tank is circulated to any of five hot water coils as called for by the thermostats. This produces five individually controlled zones.
5. Mixed fresh and recirculated air from the central air flow system is heated through the hot water coils and distributed through the five zones into rooms.
6. Cooling for summer is accomplished through a refrigeration system with a roof mounted condensing unit. The cooling coil supplies cooled air throughout the building. Hot solar heated water is introduced into the five zones to vary the temperature as required.
7. When long periods of time pass without the sun's rays (2–3 days) the water in the heat storage tank is heated with a conventional hot water boiler.

243

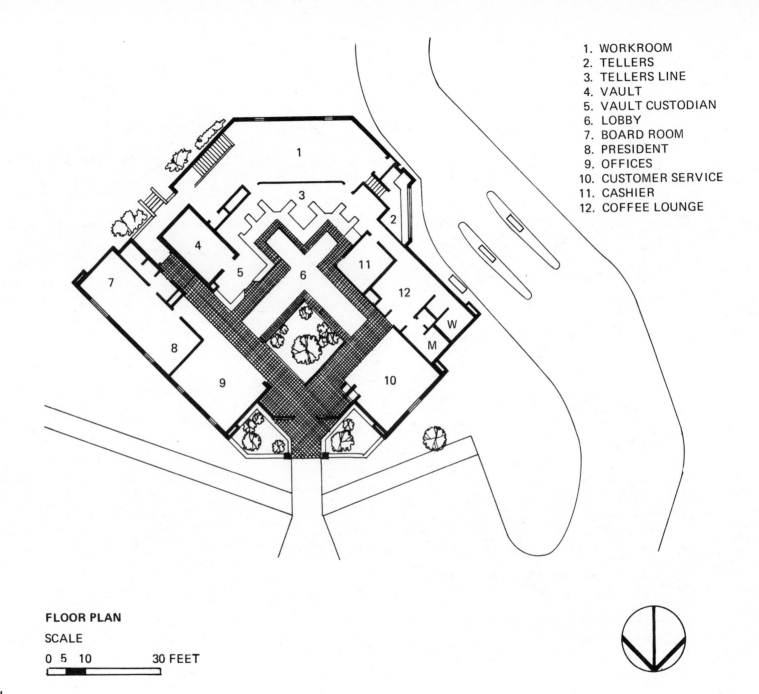

1. WORKROOM
2. TELLERS
3. TELLERS LINE
4. VAULT
5. VAULT CUSTODIAN
6. LOBBY
7. BOARD ROOM
8. PRESIDENT
9. OFFICES
10. CUSTOMER SERVICE
11. CASHIER
12. COFFEE LOUNGE

FLOOR PLAN

SCALE

0 5 10   30 FEET

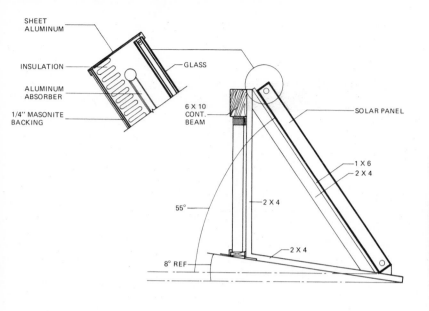

SHEET ALUMINUM

INSULATION

ALUMINUM ABSORBER

1/4" MASONITE BACKING

GLASS

6 X 10 CONT. BEAM

SOLAR PANEL

1 X 6

2 X 4

2 X 4

55°

2 X 4

8° REF

COLLECTOR SUPPORT DETAIL

Northwest elevation

Southeast elevation

Main entrance

245

Section 12. Hot Humid Region

The building presented in this section is the following:

The Frenchman's Reef Hotel. St. Thomas Island, Virgin Islands.

THE FRENCHMAN'S REEF HOTEL

LOCATION: St. Thomas Island, Virgin Islands.
LATITUDE: 18.4 degrees N.
REGION: Hot Humid.
SOLAR AND MECHANICAL ENGINEERING: Owen & Mayer
    Engineering, Lynchburg, Virginia.

MECHANICAL SOLUTION

This multi-story resort hotel has 300 rooms and a total of 250,000 square feet of air conditioned floor area. Cooling is achieved through the use of 13,000 square feet of concentrating, tracking collectors mounted on the roof of two of the hotel's wings and tilted to 6 degrees from the horizontal. At the end of the day, when the collectors have moved to their extreme position on the west, an automatic mechanism returns them to their east-facing position. The chilled water produced by the solar powered 200-ton absorption chiller is circulated through the fan-coils for space air conditioning. The heat absorbed by the chilled water is rejected to the condenser, which is operated with the sea water. The use of sea water, with its average temperature of 78 degrees F, eliminates the need for a cooling tower. The heated water from the collectors is used immediately. Therefore, there is no need for a heat storage tank.

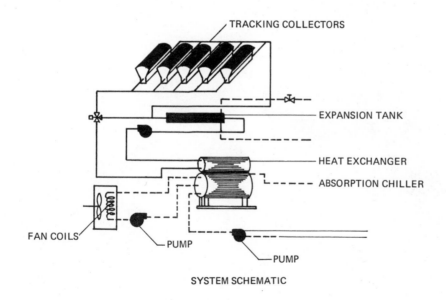

TRACKING COLLECTORS

EXPANSION TANK

HEAT EXCHANGER

ABSORPTION CHILLER

FAN COILS

PUMP

PUMP

SYSTEM SCHEMATIC

Some of the conservation improvements under way or under consideration include the capturing and recycling of rain water to reduce the load of the desalination plant, modification to room air conditioning units to operate at low or medium speeds only and recycling the condensate from desalination air ejectors.

Aerial view

Roof mounted collectors

APPENDICES

Climatic Regions of U.S.

A climatic region is a geographical area with certain defined weather characteristics. However, these characteristics are not always uniform within (or between) regions since the regions are not divided by precise lines, but instead overlap on border areas.

The general characteristics for the four climatic regions are given below.*

COOL REGION

There is limited sun during the winter months and excessive sun in the summer. Weather is sunny 58 percent of possible hours per year: in winter, 40 percent of possible hours; in summer, 74 percent. Winters are cold and long, with 48 percent of total annual hours at a temperature of 45 degrees or less. Only 4 percent of total annual hours have average temperatures above 85 degrees F. The prevailing winds come from the northwest, and over 70 percent of the time the wind causes considerable heat loss from buildings.

TEMPERATE REGION

There is a balanced distribution of clear and cloudy days. Weather is sunny 68 percent of possible hours in summer and 53 percent of possible hours during winter. Thirty-seven percent of the total annual hours are at temperatures of 45 degrees F and less. Only 6 percent of total annual hours average temperatures above 85 degrees F. Prevailing winds are from the northwest in winter and from the south during summer.

HOT ARID REGION

This region has large daily temperature variations. Sun intensity is higher here than in any other region. Weather is sunny 84 percent of possible hours year round, with 93 percent in summer and 75 percent in winter. Eleven percent of total annual hours are at temperatures of 45 degrees or less. However, 25 percent of total annual hours are at temperatures above 85 degrees F. The prevailing winds are from east to west.

HOT HUMID REGION

This region is characterized by small temperature variations. Sun intensity is high, but a large portion is cut out by clouds. Sunshine hours account for 66 percent of the possible yearly average. Eleven percent of total annual hours are at temperatures of 85 degrees F or above. The percentage of hours with temperatures of 45 degrees F or less is negligible. Winds are predominantly from the east.

---

* These standards were developed by Victor Olgay. Further information may be obtained from his book, *Design With Climate*. It is of the utmost importance to point out that these standards were developed in an effort to categorize large geographical areas. Therefore, these findings may not be applicable to specific locations, even within the described region.

Solar Energy Systems and Components

PASSIVE SOLAR SYSTEMS

Passive solar energy systems are defined as those in which only solar energy is used to transfer the thermal energy. Passive systems do not use or require a transfer medium that makes use of a secondary energy source, such as electricity, to drive a fan. The main component of a passive solar system is a large mass with a high thermal capacity that collects, stores and distributes heat.

ACTIVE SOLAR SYSTEMS

Active solar energy systems are defined as those in which energy other than solar is used to transfer the thermal energy. In the vast majority of these systems, the collection storage and distribution of the thermal energy is carried out by moving a transfer medium through the system. Active solar systems are made up of six component elements: collector, transfer, storage, distribution, back-up and controls. There are cases in which several of these elements are combined into one component.

HYBRID SOLAR SYSTEM

Hybrid solar systems are defined as those in which both passive and active technologies are integrated to produce the desired results.

COLLECTOR

The collector is the element used to absorb the solar radiation. It converts this radiation into thermal energy and transfers it to the heat removal medium. Collectors can be categorized as non-focusing and focusing. "Non-focusing" usually refers to passive or flat plate elements, and focusing collectors can be concentrating and/or tracking.

TRANSPORT

The transfer medium is the heat removing matter used in relocating the thermal energy from the collector to the storage element. The transfer medium can be air, water, water/antifreeze solution or other heat absorbing chemicals.

STORAGE

The storage element of the solar system is where the collected thermal energy is stored for later use.

DISTRIBUTION

The distribution element is responsible for the transfer of heat from the storage element to the location where it is to be used.

BACK-UP

The back-up element provides heat when the stored thermal heat is no longer available at the needed temperature or intensity. This component usually runs on conventional fuels.

MODES OF OPERATION

The sequence in which the collected thermal heat moves from the collector to its final destination varies in direct response to the system's capabilities and the user's needs. The following are several of the operating modes an active system can have.

COLLECTOR TO SPACE

The collected heat is directed from the collector to the space where it is to be used. The heat does not pass through the storage element of the system.

COLLECTOR TO STORAGE

The collected heat is directed from the collector to the storage element, where it will be used later.

AUXILIARY TO SPACE

When stored heat is inadequate for space heating, the back-up or auxiliary element can be used to provide the required thermal energy directly to the living spaces.

STORAGE TO AUXILIARY TO SPACE

When the storage contains sufficient heat but is it not at the required temperature, an energy boost is introduced prior to the usage of the stored heat.

AUXILIARY TO STORAGE TO SPACE

In cases where the storage element also serves as the distribution element, the auxiliary element will heat the storage. This allows for the storing of thermal energy produced by the usage of electricity during off-peak hours.

APPENDIX C

Supportive Organizations

The following is a list of organizations, groups, associations and government agencies that provide assistance and further information concerning solar energy and energy conservation.

American Society of Heating, Refrigerating and Air-Conditioning Engineering. 346 East 47th Street, New York, New York.

American Institute of Architects Research Corporation. 1735 New York Avenue, Washington, D.C. 20006.

Boulder Solar Energy Society. P.O. Box 3431, Boulder, Colorado 80303.

Center for Environment and Man. 275 Windsor Street, Hartford, Connecticut 06120.

Center for Research, Acts of Man. 4025 Chestnut Street, Philadelphia, Pennsylvania 19104.

Center for Science in the Public Interest. 1785 Massachusetts Avenue, Washington, D.C. 20036.

Citizens for Energy Conservation and Solar Development. P.O. Box 49173, Los Angeles, California 90049.

Department of Energy. 20 Massachusetts Avenue N.W., Washington, D.C. 20545.

Environmental Energy Foundation. 1200 Edgewater Drive, Lakewood, Ohio 44107.

Florida Solar Energy Center. 3000 State Road, 401 Cape Canaveral, Florida 32920.

HUD (Department of Housing and Urban Development), Energy Affairs. 451 7th Street S.W., Washington, D.C. 20410.

INFORM. 25 Broad Street, New York, New York 10004.

Library of Congress, Division of Science & Technology. Washington, D.C. 20540.

Mother Earth News, Inc. P.O. Box 70, Hendersonville, North Carolina 28739.

National Academy of Sciences, Building Research Advisory Board. 2101 Constitution Avenue N.W., Washington, D.C. 20418.

National Association of Home Builders. P.O. Box 1627, Rockville, Maryland 20850.

National Science Foundation, Advanced Energy Research & Technologies. Washington, D.C. 20550.

National Solar Heating and Cooling Information Center. P.O. Box 1607, Rockville, Maryland 20850. Call toll free (800) 523-2929; in Pennsylvania (800) 462-4983.

New England Sierra Club. 14 Beacon Street, Boston, Massachusetts 02108.

New England Solar Energy Association. P.O. Box 121, Townsend, Vermont 05353.

New Mexico Solar Energy Association. 6021 Canyon Road, Santa Fe, New Mexico 87501.

Solar Energy Research and Information Center. 1001 Connecticut Avenue N.W., Washington, D.C. 20036.

Solar Energy Society of America. 2780 Sepula Boulevard, Torrance, California 90510.

U.S. Weather Bureau. Federal Building, Asheville, North Carolina 28801.

Solar Equipment Mfg.

Glossary of Descriptor Codes Used in Manufacturers' and Distributors' Lists

This appendix presents an alphabetical and state listing of 558 solar equipment manufacturers. The list was compiled by the National Solar Heating and Cooling Information Center. The Center, the authors and the publishers do not endorse, recommend or attest to the quality or capability of any companies, products or individuals here listed.

NOTE: These codes were devised to cover general categories of equipment and accessories. The manufacturer or distributor should be contacted for more specific product information.

APPLICATIONS

AHW: Domestic Hot Water
AHE: Space Heating
APO: Swimming Pool Heating
ACO: Space Cooling
  ACA: Absorption Cooling
  ACD: Dessicant Cooling
AHC: Heating/Cooling
AGH: Greenhouses
AMH: Mobile Homes
AAG: Agriculture

COLLECTORS

CFA: Air Flat Plate Collectors
CFL: Liquid Flat Plate Collectors
CTU: Tubular Collectors
CCN: Concentrating Collectors
CFE: Flat Plate Evacuated Collectors
CCM: Collector Components

CMG: Glazing
CMA: Absorber Plates
CMH: Heat Transfer Fluids
CMI: Corrosion Inhibitors
CMS: Sealants
CMM: Mounting Systems
CMT: Tracking Devices
CMF: Fresnel Lenses
CMC: Selective Coatings
CMR: Reflective Surfaces
CMP: Pumps
CMB: Fans

SOLAR CONTROLS

COS: Complete Systems
COC: Components

STORAGE

STA: Air (Rock Bed)
STL: Liquid
STP: Phase Change
STC: Components

COMPLETE SYSTEMS

SYA: Air
SYL: Liquid
SYP: Passive
SSF: Solar Furnaces
SPC: Swimming Pool Covers

MISCELLANEOUS

INS: Instrumentation/Measurement
HEX: Heat Exchangers
HEP: Heat Pumps
RCE: Rankine Cycle Engine
HEL: Heliostats
KIT: Do-it-Yourself Kits
FEA: Listed in FEA Survey of Solar Collector Manufacturers
LAB: Testing Equipment
ASH: Tested According to ASHRAE Standards

ALPHABETICAL LISTING

1. A to Z Solar Products
   200 E. 26th St.
   Minneapolis, MN 55404
   (612) 870-1323
   Product Codes:
   AHW   CCM   COC   STC
   SYL   INS   KIT

2. AAI
   P.O. Box 6767
   Baltimore, MD 21204
   (301) 666-1400
   Product Codes:
   AHC   CCN   FEA

3. ABC Solar Systems
   329 Central Ave.
   Albany, NY 12206
   (518) 482-2693
   Product Codes:
   AHW   AHE   CFA

4. Ace Solar Systems
   Rt. 1/Box 50
   Mission, TX 78572
   (512) 585-6353
   Product Codes:
   AHE   AHW   CFL   SYP

5. Acurex Aerotherm
   485 Clyde Ave.
   Mountain View, CA 94042
   (415) 964-3200
   Product Codes:
   AHC   AHW   CCN   CMT
   AAG   FEA

6. Addison Products
   Addison, MI 49220
   (517) 547-6131
   Product Codes:
   AHE   AMH   CFA   COS
   STA   SYA

7. Advance Cooler Manufacturing Corp.
   Route 146, Bradford Industrial Park
   Clifton Park, NY 12065
   (518) 371-2140
   Product Codes:
   AHW   CFL   FEA   CCM
   COC   STL   STA   SYA
   SYL   SYP   AHC

8. Advanced Energy Technology
   121 Albright
   Los Gatos, CA 95030
   (408) 866-7686
   Product Codes:
   AHW   AHE   APO   CFL
   SYL

9. Advanced Solar Energy Systems Inc.
   3440 Wilshire Blvd.
   Los Angeles, CA 90010
   (213) 383-0035
   Product Codes:
   AHC   AHW   APO   CFL

10. Aerco International Inc.
    159 Paris Ave.
    Northvale, NJ 07697
    (201) 768-2400
    Product Codes:
    HEX

11. Airtex Corp.
    2900 N. Western Ave.
    Chicago, IL 60618
    (312) 463-2500
    Product Codes:
    AHW   APO   AHC   ACA

AGH   AMH   CFL   CFE
CCM   CMG   CMA   CMM
CMC   CMP   COS   COC
STL   SYL   INS   HEX
FEA   NBS   ASH

12. Airtrol Corp.
    203 W. Hawick St.
    Rockton, IL 61072
    (815) 624-8051
    Product Codes:
    AHE   CCM   CMB

13. Albuquerque Western Solar Industries
    Inc.
    612 Commanche, N.E.
    Albuquerque, NM 87107
    (505) 345-6764
    Product Codes:
    AHC   CCN   FEA   SYL

14. Alco Plastic Products
    266 Delray Ave.
    Hanover, PA 17331
    Product Codes:
    AHE   CCM   SYP   CMG

15. All Sunpower Inc.
    10400 SW 187th St.
    Miami, FL 33157
    (305) 233-2224
    Product Codes:
    AHC   AMH   CCM   CMG
    AHW   APO   CFL   SYL
    KIT   COS   COC   STL   SPC
    INS   HEX   KIT   FEA
    NBS   LAB   ASH

16. Alpha Solarco
    1014 Vine St./Ste 2230 Kroger Bldg.
    Cincinnati, OH 45202
    (513) 621-1243
    Product Codes:
    LAB   AHE   CCN   CFE

17. Alten Associates Inc.
2594 Leghorn St.
Mountain View, CA   94043
(415) 969-6474
Product Codes:
AHW   CFL   FEA   SYL
KIT   AHC   APO

18. Altenergy
P.O. Box 695
Ben Lomond, CA   95005
(408) 336-2321
Product Codes:
AHW   APO   AHE   AAG
CTU   CCN   CFL   COS
COC   HEX   RCE   LAB

19. Aluminum Co. of America
Alcoa Bldg.
Pittsburgh, PA   15219
(412) 553-2321
Product Codes:
CCM   APO   SYL   COC
CMA   CMC   CMM   CFL
AHE   AHW   FEA

20. American Acrylic Corp.
173 Marine St.
Farmingdale, NY   11735
(516) CH9-1129
Product Codes:
AHE   AHW   APO   CCM
CMG

21. American Appliance Mfg. Corp.
P.O. Box 1956/2341 Michigan Ave.
Santa Monica, CA   90404
(213) 870-8541
Product Codes:
AHW   CFL   STL   SYL

22. American Chemet Corp.
P.O. Box 165
Deerfield, IL   60015

(312) 948-0800
Product Codes:
AHE   CCM   CMC

23. American Helio Thermal Corp/Miromit
2625 S. Sante Fe Dr.
Denver, CO   80223
(303) 778-0650
Product Codes:
AHW   CFL   FEA   APO
SYL   AHC

24. American Solar Companies Inc.
Building #4/Ford Road
Denville, NJ   07834
(201) 627-0021
Product Codes:
FEA   LAB   AHC   AGH
COC   STL   INS   HEX
HEP   KIT   AHE   AHW
APO   SYL

25. American Solar Heat Corp.
7 National Pl.
Danbury, CT   06810
(203) 792-0077
Product Codes:
AHW   AHE   COC   SYL
CFL   INS

26. American Solar King Corp.
6801 New McGregor Highway
Waco, TX   76710
(817) 776-3860
Product Codes:
AHE   AHW   CFL   FEA

27. American Solar Manufacturing
P.O. Box 194
Byron, CA   94514
(415) 634-2426
Product Codes:
AHW   APO   AHE   CFL
CCM   CMP   NBS

28. American Solar Power Inc.
715 Swann Ave.
Tampa, FL   33606
(813) 251-6946
Product Codes:
AHW   CFL   FEA

29. American Solar Systems
415 Branch St.
Arroyo Grande, CA   93420
(805) 481-1010
Product Codes:
AHW   AHC   APO   ACA
ACD   AGH   AMH   AAG
CFL   CTU   COS   SYL
LAB

30. American Solarize Inc.
P.O. Box 15
Martinsville, NJ   08836
(201) 356-3141
Product Codes:
AHE   CFA   COS   STL
STA   SYA   STP

31. American Sun Industries
3477 Old Conejo Rd./P.O. Box 263
Newbury Park, CA   91320
(805) 498-9700
Product Codes:
AHE   CFL   SYL   KIT
APO   FEA

32. Ametek Inc.
1 Spring Avenue
Hatfield, PA   19440
(215) 822-2971
Product Codes:
AHW   CFL   FEA   AHC
NBS   LAB

33. Amicks Solar Heating
375 Aspen St.
Middletown, PA   17057

(717) 944-1842
Product Codes:

AHE	AHW	APO	CFL
CTU	CCM	CMG	CMA
CMH	CMC	CMS	CMP
COC	STL	SYL	HEX

34. Amprobe Instruments
630 Merrick Rd.
Lynbrook, NY 11563
Product Codes:
INS

35. Anabil Enterprizes
525 S. Aqua Clear Drive
Mustang, OK 73064
(405) 376-3324
Product Codes:
COS CCC

36. Applied Sol Tech Inc.
P.O. Box 9111 Cabrillo Station
Long Beach, CA 90810
(213) 426-0127
Product Codes:

AHC	CCN	CCM	COS
HEX	AHW	APO	STL
SYL	CMP	CMH	

37. Aqua Solar Inc.
1234 Zacchini Ave.
Sarasota, FL 33577
(813) 366-7080
Product Codes:

APO	CTU	COS	COC
SYA	SPC	KIT	FEA
ASH			

38. Aqueduct Inc.
1934 Cotner Ave.
Los Angeles, CA 90025
(213) 477-2496
Product Codes:
COS COC

39. Arizona Aluminum
249 S. 51st Ave./Box 6736
Phoenix, AZ 85005
(602) 269-2488
Product Codes:
CCM

40. Arizona Engineering & Refrigeration
635 W. Commerce Ave.
Gilbert, AZ 85234
(602) 892-9050
Product Codes:

AHE	AHW	APO	CFL
FEA	NBS		

41. Arizona Solar Enterprises
6719 E. Holly St.
Scottsdale, AZ 85257
(602) 945-7477
Product Codes:

AHE	AHW	APO	CFL

42. Arkla Industries Inc.
P.O. Box 534
Evansville, IN 47704
(812) 424-3331
Product Codes:
ACO ACA

43. Ark-Tic-Seal System Inc.
P.O. Box 428
Butler, WI 53007
(414) 276-0711
Product Codes:
AHC SYP

44. ASG Industries
P.O. Box 929
Kingsport, TN 37662
(615) 245-0211
Product Codes:

AHW	APO	CCM	AHC
CMG			

45. Astro Solar Corp.
457 Santa Anna Dr.
Palm Springs, FL 33460
(305) 965-0606
Product Codes:

AHW	AHC	ACA	AMH
AAG	CFL	CFA	CTU
CFE	COC	HEX	FEA
NBS	LAB		

46. Astron Solar Industries Inc.
465 McCormick St.
San Leandro, CA 94577
(415) 632-5400
Product Codes:

AHE	AHW	APO	CFL
COS	SYL	FEA	

47. Atlas Vinyl Products
7002 Beaver Dam Road
Levittown, PA 19057
(215) 946-3620
Product Codes:
APO CFL FEA

48. Aztec Solar Co.
P.O. Box 272
Maitland, FL 32751
(305) 628-5004
Product Codes:

AHW	CFL	CCM	COC
SYL	COS	CMM	CMP

49. Aztec Solar Energy Systems of Orange
Cty.
420 Terminal St.
Santa Ana, CA 92701
(714) 558-0882
Product Codes:
APO CFL

50. A-1 Prototype
1288 Fayette
El Cajon, CA 92020

(714) 449-6726
Product Codes:
AHE   AHW   APO   CFL
STL   SYL   KIT   FEA   NBS
ASH   COS

51. A/C Fabricating Corp.
P.O. Box 774/64600 U.S. 33 East
Goshen, IN   46525
(219) 534-1415
Product Codes:
CCM

52. Baker Bros. Solar Collectors
207 Cortez Avenue
Davis, CA   95616
(916) 756-4558
Product Codes:
AHE   AHW   APO   SYL
CFL   KIT

53. Barber-Nichols Engineering
6325 W. 55th Ave.
Arvada, CO   80002
(303) 421-8111
Product Codes:
ACO   RCE

54. Barnes Engineering Co.
30 Commerce Rd.
Stamford, CT   06904
(203) 348-5381
Product Codes:
INS

55. Berry Solar Products
Woodbridge at Main/P.O. Box 327
Edison, NJ   08817
(201) 549-3800
Product Codes:
CCM   CMC

56. Beutels Solar Heating Company
7161 N.W. 74th Street

Miami, FL   33166
(305) 885-0122
Product Codes:
AHW   CFL   FEA

57. Bi-Hex Company
P.O. Box 312
Bedford, NY   10506
(914) 764-4021
Product Codes:
COS   COC

58. Brown Manufacturing Co.
P.O. Box 14546
Oklahoma City, OK   73114
(405) 751-1323
Product Codes:
AHW.   COS   SYP   AHC
COC

59. D. W. Browning Contracting Co.
475 Carswell Avenue
Holly Hill, FL   32017
(904) 252-1528
Product Codes:
AHE   AHW   CFL   FEA
APO   SYL

60. Burke Industries Inc.
2250 South 10th Street
San Jose, CA   95112
(408) 297-3500
Product Codes:
COS   COC   KIT   APO
CFL   FEA

61. Burling Instrument Corp.
P.O. Box 298
Chatham, NJ   07928
(201) 635-9481
Product Codes:
AHW   AHE   INS

62. Butler Vent-A-Matic Corp.
P.O. Box 728
Mineral Wells, TX   76067
(800) 433-1626
Product Codes:
APO   AHW   SYL   AHE
CFL   HEX

63. C & C Solarthermics
P.O. Box 144
Smithburg, MD   21783
(301) 631-1361
Product Codes:
AHE   SSF

64. C & M Systems Inc.
P.O. Box 475/Saybrook Ind. Park
Old Saybrook, CT   06475
(203) 388-3429
Product Codes:
COC

65. California Solar Systems Co.
421 Picadilly/Suite 12
San Bruno, CA   94066
(415) 583-4711
Product Codes:
AHW   CFL   SYL

66. Calmac Manufacturing Corp.
P.O. Box 710E
Englewood, NJ   07631
(201) 569-0420
Product Codes:
AHE   AHW   CFL   FEA
APO   STP   KIT

67. Capital Solar Heating Inc.
376 N.W. 25th Street
Miami, FL   33127
(305) 576-2380
Product Codes:
AHW   APO   AHC   CFL
STL   SYL   KIT   FEA

68. Carlisle Tile & Rubber Co.
P.O. Box 99
Carlisle, PA 17013
(717) 249-1000
Product Codes:
AHW   AHE   STC

69. Carolina Aluminum
State Rd. 1184/P.O. Box 2437
Burlington, NC 27215
(919) 227-8826
Product Codes:
CCM

70. Carolina Aluminum
Metcalf Rd./Box 177
Winton, NC 27986
(919) 358-5811
Product Codes:
CCM

71. Carolina Solar Equipment Co.
P.O. Box 2068
Salisbury, NC 28144
(704) 637-1243
Product Codes:
AHE   AHW   CFL   FEA
SYL

72. Carolina Thermal Co.
Iron Works Rd./Rt. 2/Box 39
Reidsville, NC 27320
(919) 342-0352
Product Codes:
AGH   CFA

73. J. W. Carroll & Sons
22600 South Bonita St.
Carson, CA 90745
(213) 775-6737
Product Codes:
CCM   CMG

74. Castor Development Corp.
634 Crest Drive
El Cajon, CA 92021
(714) 280-6660
Product Codes:
AHE   AHW   APO   AGH
AMH   CFL   CCM   CMA
COS   COC   STL   SYL
HEX   HEP   HEL   KIT
FEA   NBS   LAB

75. Catalano & Sons Inc.
301 Stagg St.
Brooklyn, NY 11206
(212) 821-6100
Product Codes:
AHE   AHW   CFL   SYL
COC

76. Catel Mfg. Inc.
235 West Maple Ave.
Monrovia, CA 91016
(213) 359-2593
Product Codes:
APO   SPC

77. Chamberlain Manufacturing Corp.
845 Larch Avenue
Elmhurst, IL 60126
(312) 279-3600
Product Codes:
AHE   AHW   CFL   FEA

78. Champion Home Builders
5573 E. North Street
Dreyden, MI 48428
(313) 796-2211
Product Codes:
AHE   SSF   AMH   FEA

79. Chemical Processors Inc.
P.O. Box 10636
St. Petersburg, FL 33733

(813) 822-3689
Product Codes:
AHE   AHW   CFL   FEA
SYL

80. Chemical Sealing Corp.
5401 Banks Ave.
Kansas City, MO 64130
(816) 923-8812
Product Codes:
CCM   CMS

81. Chicago Solar Corp.
1773 California St.
Rolling Meadows, IL 60008
(312) 358-1918
Product Codes:
AHW   CFA

82. Clark's Products & Services
Route 1/P.O. Box 2138
Bluemont, VA 22012
(703) 955-3837
Product Codes:
AHW   AHE   APO   CFL
KIT

83. Climatrol Corp.
Woodbridge Ave.
Edison, NJ 08812
(201) 549-7200
Product Codes:
AHC   AMH   CFL   COS
COC   STL   SYL   HEX
HEP   NBS   LAB   ASH

84. Coating Laboratories
505 S. Quaker
Tulsa, OK 74120
(918) 272-1191
Product Codes:
CCM   CMR

85. Cole Solar Systems Inc.
    440A East St. Elmo Rd.
    Austin, TX 78745
    (512) 444-2565
    Product Codes:
    AHW   APO   CFL   SYL
    CCM   CMM   FEA

86. Colorado Sunworks
    P.O. Box 455
    Boulder, CO 80306
    (303) 443-9199
    Product Codes:
    AHW   AHE   AGH   AAG
    CFL   COS   COC   STL
    SYL   HEX   KIT   NBS
    LAB   ASH

87. Colt Inc.
    71590 San Jainto
    Rancho Mirage, CA 92270
    (714) 346-8033
    Product Codes:
    AHW   AHE   APO   CFL
    CCM   CMH   COS   COC
    STL   SYL   INS   HEX
    HEP   KIT   LAB   ASH

88. Columbia Technical Corp./Solar Div.
    55 High Street
    Holbrook, MA 02343
    (617) 767-0513
    Product Codes:
    AHW   APO   CFL   FEA
    CFA   COS   STL   SYA
    SYL   HEX   AHC

89. Conkling Laboratories
    5432 Merrick Road
    Massapequa, NY 11758
    (516) 541-1323
    Product Codes:
    INS

90. Conserdyne Corp.
    4437 San Fernando Road
    Glendale, CA 91204
    (213) 246-8408
    Product Codes:
    AHW   AHE   APO   CFL
    SYL   HEX

91. Consumer Energy Corporation
    4234 S.W. 75th Avenue
    Miami, FL 33155
    (305) 266-0124
    Product Codes:
    AHW   CLF   FEA   APO
    STL

92. Contemporary Systems Inc.
    68 Charlonne Street
    Jaffrey, NH 03452
    (603) 532-7972
    Product Codes:
    AHE   CFA   FEA   COS
    STA   SYA

93. Corillium Corp./Cronagold Div.
    Reston International Center
    Reston, VA 22090
    (703) 860-2100
    Product Codes:
    AHE   CCM   CMC

94. Creighton Solar Concepts
    662 Whitehead Rd.
    Lawrenceville, NJ 08648
    (609) 587-6527
    Product Codes:
    AHW   CFL

95. CSI Solar Systems Division
    12400 49th Street
    Clearwater, FL 33520
    (813) 577-4228
    Product Codes:
    AHE   AHW   CFL   FEA
    APO   COS   STL   SYL

96. Cushing Instruments
    7911 Hershel Ave.
    La Jolla, CA 92037
    (714) 459-3433
    Product Codes:
    INS

97. Cy/Ro Industries
    Wayne, NJ 07470
    (201) 839-4800
    Product Codes:
    AGH   CCM   CMG   SYP

98. C.B.M. Manufacturing Inc.
    621 NW 6th Ave.
    Fort Lauderdale, FL 33311
    (305) 463-5810
    Product Codes:
    AHW   CFL   COS   COC
    STL   SYL   HEX   FEA
    NBS   LAB

99. Daystar Corporation
    90 Cambridge St.
    Burlington, MA 01803
    (617) 272-8460
    Product Codes:
    AHE   AHW   CFL   FEA
    SYL   COC   HEX   CMP
    CMH

100. Dearing Solar Energy Systems
     12324 Ventura Blvd./P.O. Box 1744
     Studio City, CA 91604
     (213) 769-2521
     Product Codes:
     APO   SPC   CFL

101. Decker Manufacturing/Impac Corp.
     Div.
     312 Blondeau
     Keokuk, IA 52632
     (319) 524-3304

Product Codes:
AHE   CFA   KIT   SYA

102. Deko-Labs
3860 SW Archer Rd.
Gainesville, FL   32604
(904) 372-6009
Product Codes:
COC

103. Delavan Electronics Inc.
14605 North 73rd St.
Scottsdale, AZ   85260
(602) 948-6350
Product Codes:
AHE   CCM   CMT

104. Design Works
Box 700
Telluride, CO   81435
Product Codes:
AHW   CFA   AHE   COC
KIT

105. The Dexter Corp./Midland Division
East Water St.
Waukegan, IL   60085
(312) 623-4200
Product Codes:
CCM   CMC

106. Diversified Natural Resources
8025 East Roosevelt/Suite A
Scottsdale, AZ   85257
(602) 945-2330
Product Codes:
AHC   CCN   FEA

107. Dixon Energy Systems Inc.
47 East St.
Hadley, MA   01035
(413) 584-8831
Product Codes:

AHC   AHW   APO   AGH
AMH   AAG   CFL   STL
SYL

108. Diy-Sol Inc.
P.O. Box 614
Marlboro, MA   01752
Product Codes:
APO   AHE   AHW   KIT
STA   SYA   CFA   CCM
COS   COC   STL   STC

109. Dodge Products
Box 19781
Houston, TX   77024
(713) 467-6262
Product Codes:
INS

110. Dow Chemical USA
2020 Dow Center
Midland, MI   48640
(517) 636-3993
Product Codes:
AHE   AHW   CCM   CMH
CMS   CMC   CMR   STP

111. Dumont Industries
Main St.
Monmouth, ME   04259
(207) 933-4281
Product Codes:
AHW   SYL   CFL

112. Dupont Co.
Nemours Bldg. Rm. 24751
Wilmington, DE   19898
(302) 999-3456
Product Codes:
AHW   AHE   CCM   CMG
CMH

113. Ecosol Ltd.
3382 El Camino Ave.

Sacramento, CA   95821
(916) 485-5860
Product Codes:
AHC   AHW   COS   HEP
INS

114. Ecosol Ltd.
2 W. 59th St.—17th Fl./The Plaza
New York, NY   10019
(212) 838-6170
Product Codes:
AHC   AHW   COS   HEP
INS

115. Elbart Manufacturing Co.
127 West Main St.
Millbury, MA   015727
(617) 865-9412
Product Codes:
AHW   CFL   COS   COC
CCM   CMP   STL   SYL

116. Electric Motor Repair & Service
Lake Leelanau, MI   49653
(616) 256-9558
Product Codes:
AHW   CFL   CCN   CCM
CMP   LAB

117. Energex Corp.
2302 E. Magnolia St.
Phoenix, AZ   85040
(602) 267-9474
Product Codes:
AHW   AHE   APO   CFL
SYL   STL   NBS

118. The Energy Factory
5622 E. Westover/Suite 105
Fresno, CA   93727
(209) 292-6622
Product Codes:
AGH

119. Energy Absorption Systems Inc.
860 S. River Rd.
West Sacramento, CA 95691
(916) 371-3900
Product Codes:
AHW AHE CFL CCM
CMM STL SYL HEX

120. Energy Applications Inc.
Route 5/P.O. Box 383
Rutherfordton, NC 28139
(704) 287-2195
Product Codes:
AHC CCN CCM CMT
COC CMM

121. Energy Converters Inc.
2501 N. Orchard Knob Avenue
Chattanooga, TN 37406
(615) 624-2608
Product Codes:
AHE AHW CFL FEA

122. Energy Design Corp.
Box 34294
Memphis, TN 38134
(901) 382-3000
Product Codes:
AHW CCN AHC

123. Energy Dynamics Corporation
327 West Vermijo Road
Colorado Springs, CO 80903
(303) 475-0332
Product Codes:
AHW CFL FEA APO
CCM COS COC STL
STC SYL HEP AHC

124. Energy Engineering Inc.
P.O. Box 1156
Tuscaloosa, AL 35401
(205) 339-5598
Product Codes:
STL

125. Energy King
Box 248
Creston, IA 50801
(515) 782-8566
Product Codes:
AHE CFA

126. Energy Systems Inc.
4570 Alvarado Canyon Rd. Bldg. D
San Diego, CA 92120
(714) 280-6660
Product Codes:
AHW AHE CFL FEA
APO COC STL CMP
CCM CMG CMA CCN

127. Energy Systems Products Inc.
12th & Market Streets
Lemoyne, PA 17043
(717) 761-8130
Product Codes:
CCM CMM CMH

128. Enthone Inc./Sunworks Div.
P.O. Box 1004
New Haven, CT 06508
(203) 934-6301
Product Codes:
AHW CFA CFL FEA
CCM SYA SYL CMH
AHE KIT

129. Entropy Ltd.
5735 Arapahoe Ave.
Boulder, CO 80303
(303) 443-5103
Product Codes:
AHW AHC SYL STL
AAG CCN

130. Environmental Energies Inc.
Box 73 Front St.
Copemish, MI 49625
(616) 378-2000

Product Codes:
APO CFL

131. Enviropane Inc.
350 N. Marshall Street
Lancaster, PA 17602
(717) 299-3737
Product Codes:
AHE AHW CFL FEA
CFA CCM CMA

132. Eppley Laboratory Inc.
12 Sheffield Ave.
Newport, RI 02840
(401) 847-1020
Product Codes:
INS

133. Era Del Sol
5960 Mandarin Ave.
Goleta, CA 93017
(805) 967-2116
Product Codes:
AHE AHW APO CFL
SYL FEA

134. Exxon Company USA
P.O. Box 2180
Houston, TX 77001
(713) 656-0370
Product Codes:
AHC AHW CCM CMH

135. E&K Service Company
16824 74th Avenue N.E.
Bothell, WA 98011
(206) 486-6660
Product Codes:
AHE AHW CFL FEA
SYL

136. Fafco
138 Jefferson Drive
Menlo Park, CA 94025

(415) 321-6311
Product Codes:
APO   CFL   FEA   COS
SYL   HEX

137. Falbel Energy Systems Corp.
P.O. Box 6
Greenwich, CT 06830
(203) 357-0626
Product Codes:
CCN   FEA   AHW   APO
KIT   COS   CCM   CMA
CFL   STL   SYL   HEX
AHE   CMP   CMG   CMH
CMI

138. Federal Energy Corp.
5505 E. Evans
Denver, CO 80222
(303) 753-0565
Product Codes:
AHE   AHW   CFL   COC
HEX

139. Fiber-Rite Products
P.O. Box 9295
Cleveland, OH 44138
(216) 228-2921
Product Codes:
STL   STC   KIT   AHW
AHE

140. Filon
12333 S. Van Ness Ave.
Hawthorne, CA 90250
(213) 757-5141
Product Codes:
CCM   CMG

141. First International Corp.
1354 Ford St.
Colorado Springs, CO 80915
(303) 574-4404
Product Codes:

AHE   AHW   CFA   SYA
STA   COS

142. Flagala Corporation
9700 W. Highway 98
Panama City, FL 32401
(904) 234-6559
Product Codes:
AHE   AHW   CFL   FEA
APO   COS   STL

143. Florida Solar Power Inc.
P.O. Box 5846
Tallahassee, FL 32301
(904) 224-8270
Product Codes:
AHE   AHW   CFL   FEA
COC   STL   STC   HEX
APO   CCM   COS   SYL
CMP   CMA   CMG

144. Floscan Instrument Co.
3016 NE Blakely St.
Seattle, WA 98105
(206) 524-6625
Product Codes:
INS

145. Foam Products Inc.
Gay St.
York Haven, PA 17370
(717) 266-3671
Product Codes:
AHE   CFA   STA   SSF
FEA   NBS

146. Ford Products Corp.
Ford Products Rd.
Valley Cottage, NY 10989
(914) 358-8282
Product Codes:
AHW   STL

147. Fred Rice Productions
P.O. Box 643—48780 Eisenhower Dr.
La Quinta, CA 92253
Product Codes:
AHW   APO   CFL   STL
SYL   CFA   CTU   SYP
AHE   AMH

148. Calvin T. Frerichs
Chestnut Hill Rd.
Groton, MA 01450
(617) 448-6689
Product Codes:
AHW   APO   CFL   CTU
COS   COC   STL   SYL   INS
KIT   LAB

149. Friedrich Air Cond./Refrigeration
P.O. Box 1540
San Antonio, TX 78295
(512) 225-2000
Product Codes:
AHE   HEP

150. FTA Corp.
348 Hazard Ave.
Enfield, CT 06082
(203) 749-7054
Product Codes:
AHW   AHE   CFA   COS

151. Future Systems Inc.
12500 W. Cedar Dr.
Lakewood, CO 80228
(303) 989-0431
Product Codes:
AHE   APO   CFA   SSF

152. Garden Way Laboratories
P.O. Box 66
Charlotte, VT 05445
(802) 425-2147
Product Codes:
SYP   AGH   AHE

153. General Electric Co.
P.O. Box 13601/Bldg. #7
Philadelphia, PA   19101
(215) 962-2112
Product Codes:
AHW   AHC   CTU   SYL
HEP

154. General Energy Devices
1753 Ensley
Clearwater, FL   33516
(813) 586-3585
Product Codes:
AHE   AHW   CFL   FEA
APO   CCM   COS   COC
STL   SYL   CMP   HEX
CMH   NBS   ASH

155. General Solar Corp.
5575 S. Sycamore St.
Littleton, CO   80120
(303) 321-2675
Product Codes:
AHW   AHE   SYL   HEP

156. General Solar Systems
4040 Lake Park Rd./Box 2687
Youngstown, OH   44507
(216) 783-0270
Product Codes:
AHE   CCN   CCM   CMM

157. General Solargenic Corp.
P.O. Box 307
Johns Island, SC   29455
(803) 747-4480
Product Codes:
AHC   AHW   CFA   COS
STA   SYA   KIT

158. Glass-Lined Water Heater Co.
13000 Athens Ave.
Cleveland, OH   44107

(216) 521-1377
Product Codes:
AHW   STL

159. Green Mountain Homes
Royalton, VT   05068
(802) 763-8384
Product Codes:
AHC   SYP   HYB

160. Greenhouse Systems Corp.
P.O. Box 31407
Dallas, TX   75231
(214) 352-6174
Product Codes:
AHE   AAG   CBL

161. Grow House Corp.
5881 Prestonview, Suite 156
Dallas, TX   75240
Product Codes:
CCM   SYP   CMG   AGH

162. Grumman Corp./Energy Sys. Div./
Dept. G-R
4175 Veterans Memorial Highway
Ronkonkoma, NY   11779
(516) 575-6205
Product Codes:
AHE   AHW   CFL   FEA
SYL   APO   CCM   CMH
CMP

163. Grundfos Pumps Corp.
2555 Clovis Ave.
Clovis, CA   93612
(209) 299-9741
Product Codes:
CCM   CMP

164. Gulf Thermal Corporation
P.O. Box 13124 Airgate Branch
Sarasota, FL   33578
(813) 355-9783

Product Codes:
AHE   AHW   CFL   FEA
APO   SYL

165. G.N.S.
79 Magazine St.
Boston, MA   02119
(617) 442-1000
Product Codes:
APO   AHC   AAG   CCN
CCM   CMA   CMR   COS
COC   STL   SYL   INS   KIT
NBS   LAB   ASH   AHW

166. Habitat 2000 Inc.
P.O. Box 188
Belmont, NC   28012
(704) 825-5357
Product Codes:
AHW   AHE   APO   AGH
STL   SYL   HEX   KIT

167. Hadbar/Div. of Purosil Inc.
723 S. Fremont Ave.
Alhambra, CA   91803
(213) 283-0721
Product Codes:
CCM   CMS

168. Halstead and Mitchell
P.O. Box 1110
Scottsboro, AL   35768
(205) 259-1212
Product Codes:
AHE   AHW   CFL   FEA

169. Hampshire Controls Corp.
Drawer M
Exeter, NH   03833
(603) 772-5442
Product Codes:
COS   COC

170. Hansberger Refrigeration & Electric Co.
2450 8th Street
Yuma, AZ 85364
(602) 783-3331
Product Codes:
AHE   AHW   CFL   FEA

171. Harness the Sun
P.O. Box 109
Cardiff-By-The-Sea, CA 92007
(714) 436-4822
Product Codes:
APO   CFL   SYL   AHE
AHW

172. Hawthorne Industries Inc.
1501 S. Dixie
West Palm Beach, FL 33401
(305) 659-5400
Product Codes:
COS

173. HCH Associates Inc.
P.O. Box 87
Robbinsville, NJ 08691
(609) 259-9722
Product Codes:
AHE   AHW   CCM   CMG

174. Heilemann Electric
127 Mountain View Rd.
Warren, NJ 07060
(201) 757-4507
Product Codes:
CFL   HEX   STL   CTU
COC   AHW   AHE   APO

175. Helio Thermics Inc.
110 Laurens Rd.
Greenville, SC 29601
(803) 235-8529
Product Codes:
AHE   SYA

176. Heliodyne Corp.
770 S. 16th
Richmond, CA 94804
(415) 237-9614
Product Codes:
SYL   APO   AHW   AHE
CFL   KIT   CCM   CMH
COS   HEX

177. Heliodyne Inc.
4571 Linview Dr.
Rockford, IL 61109
(815) 874-6841
Product Codes:
AHC   CFA   CCM   CMA
CMM   COS   STA   SYA

178. Helics Corp.
2120 Angus Rd.
Charlottesville, VA 22901
(804) 977-3719
Product Codes:
AHC   AHW   SYA   SYL
FEA

179. Helios Solar Engineering
400 Warrington Ave.
Redwood City, CA 94063
(415) 369-6414
Product Codes:
AHW   APO   CFL   COS
STL

180. Heliotherm Inc.
W. Lenni Rd.
Lenni, PA 19052
(215) 459-9030
Product Codes:
AHE   AHW   CFL   FEA
SYL

181. Heliotrope General
3733 Kenora Drive
Spring Valley, CA 92077

(714) 460-3930
Product Codes:
AHC   AHW   APO   COC
STL   CCM   CMP   HEX

182. Helio-Dynamics Inc.
327 N. Fremont St.
Los Angeles, CA 90012
(213) 624-5888
Product Codes:
AHC   AHW   APO   CFL
SYL   FEA

183. Hercules Inc.
910 Market St.
Wilmington, DE 19899
Product Codes:
CCM   SYP   CMG   AGH

184. Hexcel Corporation
11711 Dublin Blvd.
Dublin, CA 94566
(415) 828-4200
Product Codes:
AHC   CCN   FEA

185. Highland Manufacturing Co.
P.O. Box 563
Yucaipa, CA 92399
Product Codes:
APO   CFL

186. Highland Plating Co.
1128 North Highland Ave.
Los Angeles, CA 90038
(213) 469-2288
Product Codes:
CCM   CMC

187. Hill Bros. Inc./Thermill Div.
3501 N.W. 60th Street
Miami, FL 33142
Product Codes:

AHW   AHE   APO   CFL
CCM   COS   STL   SYL
CMP   CMM

188. Hitachi Chemical Co., Ltd.
437 Madison Ave.
New York, NY   10022
(212) 838-4804
Product Codes:
AHW   APO   CFL   STL
SYL   COC   CMP   CCM

189. Hollis Observatory
One Pine St.
Nashua, NH   03060
(603) 882-5017
Product Codes:
INS

190. Honeywell Inc.
2600 Ridgeway Rd.
Minneapolis, MN   55413
Product Codes:
AHC   AHW   APO   CFL
COS   COC

191. Hudson Valley Solar
Box 388/Rt. 9
Valatie, NY   12184
(518) 781-4152
Product Codes:
AHE   SSF

192. Hydro-Flex Corp.
2101 N.W. Brickyard Rd.
Topeka, KS   66618
Product Codes:
CCM   AHC   AHW   APO

193. Ilse Engineering
7177 Arrowhead Rd.
Duluth, MN   55811
(218) 729-6858
Product Codes:

AHE   AHW   APO   CFL
CCM   CMA   STL   SYL
FEA

194. Impac Corp.
Box 365
Keokuk, IA   52632
(319) 524-3304
Product Codes:
AHE   SYA

195. In Solar Systems
2562 W. Middlefield Rd.
Mountain View, CA   94043
(415) 964-2801
Product Codes:
APO   CFL   HEX   KIT
LAB

196. Independent Energy Inc.
P.O. Box 363
Kingston, RI   02881
(401) 295-1762
Product Codes:
COC

197. Independent Living Inc.
2300 Peachford Rd.
Doraville, GA   30340
(404) 455-0927
Product Codes:
AHC   AHW   APO   COC
COS   SYL

198. International Environment Corp.
83 S. Water St.
Greenwich, CT   06830
(203) 531-4490
Product Codes:
AHW   CFL   FEA   AHE
SYL

199. International Environment Corp.
1400 Mill Creek Rd.

Gladwyne, PA   19035
(215) 642-3060
Product Codes:
AHW   CFL   FEA   AHE
SYL

200. International Environmental Energy Inc.
275 Windsor St.
Hartford, CT   06120
Product Codes:
AHE   AHW   COS   COC
SYA   SYL

201. International Solar Industries Inc.
3107 Memorial Pkwy. NW
Huntsville, AL   35801
Product Codes:
AHE   AHW   CFA   SYL

202. International Solar Industries Inc.
9555 East Caley
Englewood, CO   80110
Product Codes:
AHE   AHW   CFA   SYL

203. Intertechnology/Solar Corp. of America
100 Main St.
Warrenton, VA   22186
(703) 347-7900
Product Codes:
AHW   CFL   SYL   KIT
AHC   APO   CCN   FEA

204. Iowa Solar Electronics
P.O. Box 246
North Liberty, IA   52317
(319) 626-2342
Product Codes:
COC

205. Itek Corp-Optical Systems Div.
10 Maguire Rd.
Lexington, MA   02117
(617) 276-5825

Product Codes:

AHW	APO	CFL	CCM
CMF	CMC	CMR	FEA
NBS	LAB		

206. ITT/Fluid Handling Div.
4711 Golf Road
Skokie, IL 60076
(312) 677-4030
Product Codes:

AHC	AHW	CCM	CMP
HEX	COC		

207. J & J Solar
7273 North Central Ave.
Phoenix, AZ 85020
(602) 956-9536
Product Codes:

AHW	AHE	APO	AGH
AMH	AAG	CFL	CCM
CMA	CMH	CMM	CMC
COS	COC	STL	SYL
HEX	KIT	ASH	

208. W. L. Jackson Manufacturing Co.
1200-25 W. 40th Street
Chattanooga, TN 37401
(615) 867-4700
Product Codes:

AHE	SYL	CFL

209. Jacobs-Del Solar Systems Inc.
251 S. Lake Ave.
Pasadena, CA 91101
(213) 449-2171
Product Codes:

CCN	AHC	AHW	APO
CCM	CMT	COC	SYL
FEA			

210. Johnson Controls Inc./Penn Division
2221 Camden Ct.
Oak Brook, IL 60521

(312) 654-4900
Product Codes:

AHE	AHW	APO	COS
COC			

211. J. G. Johnston Company
33458 Angeles Forest Hwy.
Palmdale, CA 93550
(805) 947-3791
Product Codes:

AHE	CFA	FEA	AHW
SYA	STA		

212. Kalwall Corp./Solar Components Div.
Box 237
Manchester, NH 03105
(603) 668-8186
Product Codes:

AHE	APO	CFA	CFL
CCM	STL	STC	SYA
AHW	CMG	SYP	CMA
CMS	CMC	COC	COS
CMB	CMP	SYL	

213. Kasaki USA
4150 Arch Dr./Suite 8
Studio City, CA 91604
(213) 985-9611
Product Codes:

AHW	APO	CFL	COS
STL	SYL		

214. Kastek Corp.
P.O. Box 8881
Portland, OR 97208
Product Codes:

APO	AHW	AAG	CFL
SYL			

215. Kem Associates
153 East St.
New Haven, CT 06507
(203) 865-0584

Product Codes:

CCM	CMM

216. Kennecott Copper Corporation
128 Spring Street
Lexington, MA 02173
(617) 862-8268
Product Codes:

AHE	AHW	CCM	APO
CMA			

217. Kentucky Solar Energies Inc.
Rt. 1/Box 278
Frankfort, KY 40601
Product Codes:

AHW	AHE	APO	CFL	
COS	COC	STL	SYL	INS
HEX	FEA	NBS	LAB	

218. Mel Kiser and Assoc.
6701 E. Kenyon Dr.
Tucson, AZ 85710
Product Codes:

AHW	AHC	CCN
SYL		

219. Kreft Distributing Co.
Box 105
Lake Havasu City, AZ 86403
(602) 855-2059
Product Codes:

AHC	AHW	APO	CFL
CCM	COC	STL	SYL
KIT	CMP	CMG	CMA

220. KTA Corporation
12300 Washington Avenue
Rockville, MD 20852
(301) 468-2066
Product Codes:

AHE	AHW	FEA	CCN
SYL	CTU		

221. K-Line Corp.
911 Pennsylvania Ave. N.E.
Albuquerque, NM 87110
(505) 268-3379
Product Codes:

| AHE | CFA | STA | SYA |
| AHW | COS | INS | |

222. Wm. Lamb Co.
P.O. Box 4185
North Hollywood, CA 91607
(213) 764-6363
Product Codes:

AHC   COC   KIT

223. Largo Solar Systems Inc.
991 SW 40th Ave.
Plantation, FL 33317
(305) 583-8090
Product Codes:

AHW	CFL	FEA	APO
CCM	SYL	KIT	COS
CMP	CMA		

224. Lennox Industries Inc.
350 S. 12th Avenue–P.O. Box 280
Marshalltown, IA 50158
(515) 754-4011
Product Codes:

| AHE | AHW | CFL | FEA |
| SYL | HEP | | |

225. Life Star of Kansas
N. Highway 25
Atwood, KS 67730
Product Codes:

AHE   SYA

226. Lof Solar Energy Systems
1701 Broadway
Toledo, OH 43605
(419) 247-4355
Product Codes:

| AHE | AHW | CFL | FEA |
| CCM | CMG | COC | COS |

227. MacBall Industries
3040 Market St.
Oakland, CA 94608
(415) 658-2228
Product Codes:

SPC

228. Mann-Russell Electronics Inc.
1401 Thorne Rd.
Tacoma, WA 98421
(206) 383-1591
Product Codes:

| AHE | CCM | CMM | HEL |
| KIT | | | |

229. March Manufacturing Co. Inc.
1819 Pickwick Ave.
Glenview, IL 60025
(312) 729-5300
Product Codes:

CCM   CMP

230. Martin Processing Inc.
P.O. Box 5068
Martinsville, VA 24112
(703) 629-1711
Product Codes:

AGH   AAG   CCM   CMG

231. McKim Solar Energy Systems Inc.
2627 E. Admiral Place
Tulsa, OK 74110
(918) 936-4035
Product Codes:

AHC	AHW	APO	CFA	
CFL	CTU	CCN	CCM	
CMT	COS	COC	STA	
STL	STP	HEX	KIT	NBS
LAB	ASH			

232. Mechanical Mirror Works
661 Edgecombe Ave.
New York, NY 10032
(212) 795-2100
Product Codes:

AHC   CCM   CMR

233. Midwest Solar Corp.
2359 Grissom Drive
St. Louis, MO 63141
(314) 569-3110
Product Codes:

| AHW | AGH | CFA | CCN |
| STL | SYL | LAB | |

234. Mid-West Technology Inc.
P.O. Box 26238
Dayton, OH 45426
(513) 274-6020
Product Codes:

AHE   CCM   CMA

235. Mid-Western Solar Systems
2235 Irvin Cobb Dr./Box 2384
Paducah, KY 42001
(502) 443-6295
Product Codes:

| AAG | AHE | AHW | CFL |
| SSF | FEA | | |

236. Miromit/American Heliothermal
2625 S. Santa Fe Dr.
Denver, CO 80223
(303) 778-0650
Product Codes:

| AHW | CFL | FEA | APO |
| SYL | AHC | | |

237. Mor-Flo Industries Inc.
18450 South Miles Rd.
Cleveland, OH 44128
(216) 663-7300
Product Codes:

AHW   SYL   STL   CFL

KIT   CCM   CMH   CMI
CMM

238. National Energy Co.
21716 Kendrick Ave.
Lakeville, MN 55044
(612) 469-3401
Product Codes:
AHE   AHW   CFA   SYA
STA   COS   COC   NBS
FEA   CCM   CMB

239. National Energy Systems Corp.
P.O. Box 1176
Birmingham, AL 35201
(205) 252-7726
Product Codes:
AHE   AHW   APO   CFL
SYL   COS   COC   STL
ASH

240. National Industrial Sales
6501 W. 99th St.
Chicago Ridge, IL 60415
(312) 423-4924
Product Codes:
AHE   AHW   CCM   CMG

241. National Solar Company
2331 Adams Drive N.W.
Atlanta, GA 30318
(404) 352-3478
Product Codes:
AHE   AHW   CFL   FEA
SYL

242. National Solar Corp.
Novelty Lane
Essex, CT 06426
(203) 767-1644
Product Codes:
AHW   CFL   CCM   CMM
KIT

243. National Solar Sales Inc.
165 West Wieuca Road/Suite 100
Atlanta, GA 30342
(404) 256-1660
Product Codes:
AHW   AHC   APO   CFL
STL   SYL

244. National Solar Systems
P.O. Box 82177
Tampa, FL 33682
(813) 935-9634
Product Codes:
AHW   AHE   APO   CFL
COS   COC   STL   SYL
HEP   KIT   FEA   NBS

245. Natural Energy Corp.
1001 Connecticut Ave. NW
Washington, DC 20036
(202) 296-7070
Product Codes:
AHW   AHC   APO   CFL
AAG

246. Natural Energy Systems Inc.
1117 E. Carpenter Dr.
Palatine, IL 60067
(312) 359-6760
Product Codes:
AHW   AHC   APO   CFL
COS   STL   HEX   HEP
SYL

247. Natural Energy Systems/Marketing
Arms Dv.
1632 Pioneer Way
El Cajon, CA 92020
(714) 440-6411
Product Codes:
AHW   CFL   FEA   APO
AHC

248. Natural Heating Systems
207 Cortez Ave.
Davis, CA 95616
(916) 756-4558
Product Codes:
AHW   AHE   APO   CFL

249. Natural Power Inc.
Francestown Turnpike
New Boston, NH 03070
(603) 487-5512
Product Codes:
AHE   COS   COC

250. New Jersey Aluminum/Solar Div.
1007 Jersey Ave./P.O. Box 73
North Brunswick, NJ 08902
(201) 249-6867
Product Codes:
CCM   CMA   KIT   AHE
AHW

251. Northern Solar Power Co.
311 S. Elm St.
Moorhead, MN 56560
(218) 233-2515
Product Codes:
AHW   AHE   CFL

252. Northrup Inc.
302 Nichols Drive
Hutchins, TX 75141
(214) 225-4291
Product Codes:
AHC   AHW   CFL   CCN
FEA   APO

253. Northwest Solar Systems Inc.
7700 12th NE
Seattle, WA 98115
(206) 523-3951
Product Codes:
AHE   AHW   APO   CFA
COC   STA   SYA

254. M. C. Nottingham Co.
4922 Irwindale Ave./P.O. Box 2107
Irwindale, CA   91706
(213) 283-0407
Product Codes:
AHE   STL

255. NRG Ltd.
901 2nd Ave. East
Coralville, IA   52241
(319) 354-2033
Product Codes:
AHE   CCN   AHW

256. NRG Manufacturing
P.O. Box 53
Napoleon, OH   43545
(419) 599-3618
Product Codes:
AHE   SSF   FEA

257. Nuclear Technology Corp.
P.O. Box 1
Amston, CT   06231
(203) 537-2387
Product Codes:
AHE   AHW   CCM   CMH
CMI

258. Ocli—Optical Coating Lab Inc.
2789 Giffen Ave.
Santa Rosa, CA   95403
(707) 545-6440
Product Codes:
CCM   CMG   CMA   CMC
CMR

259. Odin Solar Systems
26010 Eden Landing Rd/Suite 5
Hayward, CA   94545
(415) 785-2000
Product Codes:
CCM   CMA   CMP   INS

260. Oem Products Inc./Solarmatic
2413 Garden St.
Tampa, FL   33605
(813) 247-5848
Product Codes:
AHW   APO   CFL   COC
AHC   FEA   CMM   CMP
HEP

261. Ohio Valley Solar Inc.
4141 Airport Rd.
Cincinnati, OH   45226
(513) 871-1961
Product Codes:
AHE   APO   CFA   COS
STA   SYA   LAB   SSF

262. Olin Brass Corp./Roll-Bond Div.
E. Alton, IL   62024
(618) 258-2000
Product Codes:
AHW   CCM   AHC   CMA

263. One Design Inc.
Mountain Falls Route
Winchester, VA   22601
(703) 662-4898
Product Codes:
SYP

264. Optical Sciences Group Inc.
24 Tiburon St.
San Rafael, CA   94901
(415) 453-8980
Product Codes:
AHE   CCM   CMF

265. Oriel Corp. of America
15 Market St.
Stamford, CT   06902
(203) 357-1600
Product Codes:
INS   LAB

266. Overly Manufacturing Co.
574 W. Otterman St.
Greensburg, PA   15601
(412) 834-7300
Product Codes:
AHE   AHW   APO   CFL

267. Owen Enterprises
436 No. Fries Ave.
Wilmington, CA   90744
(213) 835-7436
Product Codes:
AHC   CCN   APO

268. Owens Illinois Inc.
P.O. Box 1035
Toledo, OH   43666
(419) 242-6543
Product Codes:
AHC   CTU   FEA   AHW

269. P C A
11031 Wye Drive
San Antonio, TX   78217
(512) 656-9338
Product Codes:
NBS   STL   CCM   APO
AHW   AHE   CFL   SYL
KIT   CMP

270. Packless Industries Inc.
P.O. Box 310
Mount Wolf, PA   17347
Product Codes:
AHW   HEX   APO   AHC

271. Pak-Tronics Inc.
4044 N. Rockwell Ave.
Chicago, IL   60618
(312) 478-8585
Product Codes:
AHE   AHW   COS   COC

272. Park Energy Co.
Box SR9
Jackson, WY 83001
(307) 733-4950
Product Codes:
CFA   STA   AHE   AHW

273. People/Space Co.
259 Marlboro St.
Boston, MA 02109
(617) 742-8652
Product Codes:
AHW   AHE   APO   AGH
AMH   CFL   COC   INS
KIT   LAB

274. Permaloy Corp.
P.O. Box 1559
Ogden, UT 84402
(801) 731-4303
Product Codes:
AHE   CCM   CMC

275. Pioneer Energy Products
Route 1/P.O. Box 189
Forest, VA 24551
(804) 239-9020
Product Codes:
AHW   AHC   CFL   COS
SYL   NBS

276. Piper Hydro Inc.
2895 East La Palma
Anaheim, CA 92806
(714) 630-4040
Product Codes:
AHE   AHW   CFL   FEA
APO   SYL

277. Pleiad Industries, Inc.
Springdale Rd.
West Branch, IA 52358
(319) 356-2735

Product Codes
AHE   AHW   APO   CFL
FEA

278. Powell Brothers, Inc.
5903 Firestone Blvd.
South Gate, CA 90280
(213) 869-3307
Product Codes:
AHE   AHW   CFL   FEA
APO

279. PPG Industries
One Gateway Center
Pittsburgh, PA 15222
(412) 434-3555
Product Codes:
AHE   AHW   CFL   FEA
APO

280. Practical Solar Heating
209 S. Delaware Dr. Rt. 611
Easton, PA 18042
(215) 252-6381
Product Codes:
AHE   COC   CFL   STC
APO   KIT   CCM   AHW
HEX   STL   CMP   CTU

281. Prima Industries, Inc.
P.O. Box 141
Deer Park, NY 11729
(516) 242-6347
Product Codes:
AHW   CFL   SYL

282. Professional Fiberglass Products, Inc.
Ada Industrial Park/P.O. Box1179
Ada, OK 74820
(405) 436-0223
Product Codes:
STL

283. Pyco
600 E. Lincoln Hwy.
Penndel, PA 19047
(215) 757-3704
Product Codes:
INS

284. Ranco Inc.
601 W. Fifth Ave.
Columbus, OH 43201
Product Codes:
AHE   AHW   COC

285. Raypak Inc.
31111 Agoura Rd.
Westlake Village, CA 91359
(213) 889-1500
Product Codes:
AHE   AHW   APO   CFL
FEA   SYL   CCM   CMP
NBS

286. Ra-Los Inc.
559 Union Ave.
Campell, CA 95008
(408) 371-1734
Product Codes:
INS   RCE   FEA   NBS
LAB   ASH   APO   AAG
CCM   CMG   CMH   CMI
CMM   CMP   AHC   AHW
CFL   SYL   STL   HEX
COS   CDC

287. Research Products Corp.
P.O. Box 1467/1015 E. Washington Ave.
Madison, WI 53701
(608) 257-8801
Product Codes:
AHW   AHE   CFA   CCM
CMM   CMB   COS   LAB

288. Research Technology Corp.
151 John Downey Dr.
New Britain, CT 06051
(203) 224-8155
Product Codes:
CCM   CMH   INS   HEX

289. Resource Technology
151 John Downey Dr.
New Britain, CT 06051
(203) 224-8155
Product Codes:
AHC   AHW   CCM   CMH

290. Revere Copper & Brass, Inc.
P.O. Box 151
Rome, NY 13440
(315) 338-2401
Product Codes:
AHW   CFL   FEA   APO
CCM   SYL   COS   COC
AHE   STL   CMP   NBS

291. Reynolds Metals Company
P.O. Box 27003
Richmond, VA 23261
(804) 282-3026
Product Codes:
AHE   AHW   CFL   FEA
APO   NBS

292. Rheem Water Heater Div./City
Investing
7600 S. Kedzie Ave.
Chicago, IL 60652
(312) 434-7500
Product Codes:
AHW   STL

293. Rho Sigma
11922 Valerio St.
North Hollywood, CA 91605
(213) 982-6800
Product Codes:

AHW   AHE   APO   COS
COC   INS

294. Richdel, Inc.
P.O. Drawer A/1851 Oregon St.
Carson City, NV 89701
(702) 882-6786
Product Codes:
COC   APO   AHW

295. C.F. Roark Welding & Eng'ng. Co., Inc.
136 N. Green St.
Brownsburg, IN 46112
(317) 852-3163
Product Codes:
CFL   CCN   AHE   KIT

296. W.R. Robbins & Son
1401 N.W. 20th Street
Miami, FL 33142
(305) 325-0880
Product Codes:
AHW   CFL   FEA   APO
SYL   COS   COC   KIT
CCM   CMP   CMH   STL

297. Robert Shaw Controls Co./Uni-Line
Div.
P.O. Box 2000/4190 Temescal St.
Corona, CA 91720
(714) 734-2600
Product Codes:
AHC   AHW   APO   COC

298. Rohm & Haas/Plastics Enginering
P.O. Box 219
Bristol, PA 19067
(215) 788-5501
Product Codes:
CCM   CMG   CMS

299. Rox International
2604 Hidden Lake Dr/Suite D
Sarasota, FL 33577

(813) 366-6053
Product Codes:
AHW   AHE   CCN   RCE
LAB

300. Milton Roy Co/Hartell Div
70 Industrial Dr.
Ivyland, PA 18974
(215) 322-0730
Product Codes:
CCM   CMP

301. R.M. Products
5010 Cook St.
Denver, CO 80216
(303) 825-0203
Product Codes:
AHE   AHW   CFL   CFA
SYA   FEA   CCM   CMM
APO   AGH   COS   COC
STA   STL   SYL   SYP

302. S. S. Solar Inc.
16 Keystone Ave.
River Forest, IL 60305
(312) 771-1912
Product Codes:
AHW   AHC   CCN   SYL

303. Salina Solar Products Inc.
620 N. 7th
Salina, KS 67401
(913) 823-2131
Product Codes:
AHW   CFL   STL   SYL

304. Science Associates Inc.
230 Nassau St./P.O. Box 230
Princeton, NJ 08540
(609) 924-4470
Product Codes:
INS

305. Scientific-Atlanta Inc.
3845 Pleasantdale Road
Atlanta, GA 30340
(404) 449-2000
Product Codes:
AHE   AHW   CFL   FEA

306. Sealed Air Corp.
2015 Saybrook Ave.
Commerce, CA 90040
(213) 685-9666
Product Codes:
APO   SPC

307. Seeco - Solar Engineering & Equip. Co.
3305 Metairie Rd.
Matairie, LA 70001
(504) 837-0676
Product Codes:
AHW   AHE   AAG   CFA
COS   COC   STA   SYA
NBS   ASH

308. Semo Solar Products Corp.
1054 N.E. 43rd St.
Ft. Lauderdale, FL 33334
(305) 565-2516
Product Codes:
AHE   AHW   CFL   FEA
APO   SYL   STL   COC
CCM   CMP

309. Shape Symmetry & Sun Inc.
Biddeford Industrial Park
Biddeford, ME 04005
(207) 282-6155
Product Codes:
AHW   CFL   COS   COC
STL   SYL

310. Sheldahl/Advanced Products Div.
Northfield, MN 55057
(507) 645-5633
Product Codes:
AHE   AHW   CCM   CMR

311. Shell Oil Co.
1 Shell Plaza/P.O. Box 2463
Houston, TX 77001
Product Codes:
AHC   CCM   CMH

312. Shelley Radiant Ceiling Co.
8110 N. St. Louis Ave.
Skokie, IL 60076
Product Codes:
APO   CCM

313. Sigma Energy Products
720 Rankin Rd. N.E.
Albuquerque, NM 87107
(505) 344-3431
Product Codes:
AHE   AHW   APO   CFL
FEA

314. J. & R. Simmons Construction Co.
2185 Sherwood Dr.
South Daytona, FL 32019
(904) 677-5832
Product Codes:
AHE   AHW   CFL   SYL
FEA

315. Simons Solar Environmental Systems,
Inc.
24 Carlisle Pike
Mechanicsburg, PA 17055
(717) 697-2778
Product Codes:
AHW   SYL   AHE   CFL
CCM   COS   KIT   COC
CMA   CMG   CMP   FEA

316. Skytherm Processes Engrg.
2424 Wilshire Blvd.
Los Angeles, CA 90057
(213) 389-2300
Product Codes:
AHC   AHW   SYP   FEA

317. A. O. Smith Corp.
Box 28
Kankakee, IL 60901
Product Codes:
AHW   STL   STC

318. Sol Ray
204 B Carleton
Orange, CA 92667
(714) 997-9431
Product Codes:
AHW   AHE   APO   AMH
AAG   CFA   CCM   CMG
CMA   CMI   CMS   CMM
CMR   CMP   CMB   COS
COC   STA   SYA   INS
HEX   LAB

319. Sola Heat
1200 East 1st St.
Los Angeles, CA 90033
(213) 263-5823
Product Codes:
AHE   AHW   CFL   STL
SYL

320. Solafern Ltd.
P.O. Box M
Buzzards Bay, MA 02532
(617) 759-7527
Product Codes:
AHE   AHW   CFA   COS

321. The Solar Store
P.O. Box 841
Peoria, IL 61652
(309) 673-0458
Product Codes:
AHW   AHE   AAG   CFA

322. Solar Alternative Inc.
30 Clark St.
Brattleboro, VT 05301
(802) 254-8221

Product Codes:
AHE   AHW   CFL   APO

323. Solar America Inc.
9001 Arbor St.
Omaha, NE 68124
(402) 397-2421
Product Codes:
AHC   AHW   ACA   AAG
CFL   CFA   STL   SYA
HEP

324. Solar Applications Inc.
7926 Convoy Court
San Diego, CA 92111
(714) 292-1857
Product Codes:
AHE   AHW   APO   CFL
FEA   CCM   CMM   SPC
NBS

325. Solar Applications Inc.
One Washington St.
Wellesley, MA 02181
(617) 237-5675
Product Codes:
AHW   CFL   SYL   STL

326. Solar Aqua Heater Corp.
15 Idlewell St.
Weymouth, MA 02188
(617) 843-7255
Product Codes:
CCM   CMR

327. Solar Captivators Systems Inc.
7192 Clairemont Mesa Blvd.
San Diego, CA 92111
(714) 560-7454
Product Codes:
AHW   APO   AHC   AAG
CFL

328. Solar Central
7213 Ridge Road
Mechanicsburg, OH 43044
(513) 828-1356
Product Codes:
AHW   CFL   FEA   CCM
HEP   AHC   KIT   SYL
STL   CMP   COS   AGH
ACA   SYP   RCE   CMM
CMG

329. Solar Collectors of Santa Cruz
2902 Glen Canyon Rd.
Santa Cruz, CA 95060
(408) 476-6369
Product Codes:
AHW   AHE   APO   AHC
AGH   AMH   AAG   CFA
CFL   CCN   CCM   CMA
CMM   STA   STL   SYP
SYL   SYA   KIT   HEX

330. Solar Comfort Inc.
Route 3/Box 139
Statesville, NC 28677
(704) 872-0753
Product Codes:
AHE   SSF

331. Solar Comfort Systems/Div. Solar Sys.
4853 Cordell Aven./Suite 606
Bethesda, MD 20014
(301) 652-8941
Product Codes:
AHE   AHW   APO   CFL
SYL   CFA   SYA   STL
COC   KIT   CCM   CMG
FEA

332. Solar Contact Systems
1415 Vernon
Anaheim, CA 92805
(714) 991-8120
Product Codes:

AHW   CFL   STL   SYL
KIT

333. Solar Control Corp.
P.O. Box 2201/201 W FM 2410
Harket Heights, TX 76541
(817) 699-8858
Product Codes:
AHW   AHE   CFA   CFL
CCN

334. Solar Control Corp.
5595 Arapahoe
Boulder, CO 80302
(303) 445-9180
Product Codes:
AHE   COC   COS

335. Solar Controlar Inc.
P.O. Box 8703
Orlando, FL 32806
Product Codes:
AHE   COS   AHW

336. Solar Corp. of America/Intertechnology
100 Main Street
Warrenton, VA 22186
(703) 347-7900
Product Codes:
AHE   AHW   CFL   FEA
SYL   APO   KIT

337. Solar Development Inc.
4180 Westroads Drive
West Palm Beach, FL 33407
(305) 842-8935
Product codes:
AHE   AHW   CFA   CFL
FEA   APO   CCM   COC
SYL   CMM   CMA   CMP
NBS

338. Solar Dynamics Corporation
550 Frontage Rd.

Northfield, IL 60093
(312) 446-5242
Product Codes:
AHW   CFL   SYL   AHE

339. Solar Dynamics Inc.
P.O. Box 3457
Hialeah, FL 33013
(305) 921-7911
Product Codes:
AHE   AHW   CFL   FEA
SYL

340. Solar Dynamics Inc.
1320 S. Lipan St.
Denver, CO 80223
(303) 777-3666
Product Codes:
AHE   SSF

341. Solar Electric Inc.
403 S. Maple
West Branch, IA 52329
(319) 643-2598
Product Codes:
AHW   AHE   APO   AMH
AAG   CCN   CCM   CMG
CMA   CMS   CMM   COS
COC   STA   SYA   KIT
FEA   NBS   ASH

342. Solar Electric International
2634 Taft Ave.
Orlando, FL 32804
(305) 422-8396
Product Codes:
AHC   AHW   CTU   SYL

343. Solar Energy Applications, Inc.
1102 E. Washington St.
Phoenix, AZ 85034
(602) 244-1822
Product Codes:

AHW   APO   CFL   STL
SYL   KIT   LAB

344. Solar Energy Company
P.O. Box 649
Gloucester Point, VA 23062
Product Codes:
AHE   AHW   APO   CFL

345. Solar Energy Components, Inc.
1605 North Cocoa Blvd.
Cocoa, FL 32922
(305) 632-2880
Product Codes:
AHE   AHW   CFL   FEA
CCM   CMP   CMG   KIT
COC

346. Solar Energy Contractors
P.O. Box 17094
Jacksonville, FL 32216
(904) 641-5611
Product Codes:
AHE   AHW   CFL   FEA
COS   COC   STL   KIT
CCM   CMP   CMM

347. Solar Energy Digest
P.O. Box 17776
San Diego, CA 92117
Product Codes:
AHW   CFL   KIT

348. Solar Energy Engineering
31 Maxwell Court
Santa Rosa, CA 95401
(707) 542-4498
Product Codes:
AHW   AHE   APO   CFL

349. Solar Energy Inc.
12155 Magnolia Ave/6-E
Riverside, CA 92503
(714) 785-0610

Product Codes:
AHW   AHC   APO   CFL

350. Solar Energy People
5044 Fair Grounds Drive
Mariposa, CA 95338
(209) 966-5616
Product Codes:
AHW   APO   CFL   CCM
CMA   COS   COC   STL
SYL   KIT

351. Solar Energy Products Company
121 Miller Road
Avon Lake, OH 44012
(216) 933-5000
Product Codes:
AHE   AHW   FEA   HEX
APO   SYA   COC   STA
CFA   AAG   CCM   CMB

352. Solar Energy Products Inc.
1208 N.W. 8th Ave.
Gainesville, FL 32601
(904) 377-6527
Product Codes:
AHE   AHW   APO   CFL
STL   SYL   KIT   HEX
COC   CCM   CMM   CMP
CMG   NBS

353. Solar Energy Research Corporation
7018 South Main St.
Longmont, CO 80501
(303) 772-8406
Product codes:
AHC   AHW   APO   HEP
CCM   CMM   CMP   STL
KIT   COS   COC   CMA
CMT   INS   HEX   NBS

354. Solar Energy Research & Development
302 Lucas
Mt. Pleasant, SC 29464

(803) 884-0290
Product Codes:

AHW	AHE	CFL	COS
STL	SYL		

355. Solar Energy Resources Corporation
10639 SW 185 Terrace
Miami, FL 33157
(305) 233-0711
Product Codes:

AHW	CFL	FEA	AHC
APO	SYL	HEP	ACA

356. Solar Energy Systems Inc.
One Olney Ave.
Cherry Hill, NJ 08003
(609) 424-4446
Product Codes:

AHW	CFL	FEA	COS
STL	SYL	AHC	APO
HEP			

357. Solar Energy Systems of Georgia
5825 Glenridge Dr. N.E./Bldg.2
Atlanta, GA 30328
(404) 255-9588
Product Codes:

AHW	COS	COC	SYL
CFL	SYA	STL	CCM
CMP	HEX		

358. Solar Energy Systems & Products
500 N Alley
Emmitsburg, MD 21727
(301) 447-6354
Product Codes:

AHE	CFL	AHW	COS

359. Solar Engineering & Mfg. Co. Inc.
P.O. Box 1358
Boca Raton, FL 33432
(305) 368-2456
Product Codes:

INS	HEL	KIT	FEA

NBS	LAB	ASH	AHW
APO	CFL	CCN	CCM
CMG	CMA	CCM	CMT
CMR	COS	COC	STL
SYL			

360. Solar Enterprises
9803 E. Rush Street
El Monte, CA 91733
(213) 444-2551
Product Codes:

APO	CFL	FEA

361. Solar Enterprises
P.O. Box 1046
Red Bluff, CA 96080
(916) 527-0551
Product Codes:

AHE	AHW	APO	AMH
AAG	CFL	CCM	CMG
CMA	CMM	CMR	COS
COC	STL	SYL	INS
LAB			

362. Solar Enterprises Hawaii
P.O. Box 27031
Honolulu, HI 96827
(808) 922-8528
Product Codes:

AHW	CFL	KIT

363. Solar Enterprises Inc.
7830 N Beach St.
Minneapolis, MN 55440
(612) 483-8103
Product Codes:

AHW	AHC	APO	AGH
AMH	AAG	CFL	CCM
CMG	CMA	CMM	CMC
CMP	COS	COC	STL
SYL	HEX	HEP	LAB
ASH			

364. Solar Enterprises Inc.
2816 West Division
Arlington, TX 76012
(817) 461-5571
Product Codes:

AHE	CCM	CMA	CMM
CMC	COS	COC	STL
HEX	KIT	FEA	ASH
AHW	APO	CFL	SYL

365. Solar Equipment Corp.
Woodbridge at Main/P.O. Box 327
Edison, NJ 08817
(201) 549-3800
Product Codes:

CCM	CMA	CMC

366. Solar Equipment Corp.
P.O. Box 357
Lakeside, CA 92040
Product Codes:

AHC	AHW	APO	CFL

367. Solar Farm Industries Inc.
P.O. Box 242
Stokton, KS 67669
(913) 425-6726
Product Codes:

AHW	AHE	APO	AGH
AMH	AAG	CFA	CCM
CMA	CMH	CMC	COS
COC	KIT	NBS	ASH

368. Solar Fin Systems
140 S. Dixie Highway
St. Augustine, FL 32084
(904) 824-3522
Product Codes:

AHW	AHE	CFL	COS
SYL			

369. Solar Glo-Thermal Energy Systems Inc.
P.O. Box 377
Dayton, OH 45459

(513) 252-6150
Extensive Product Line—Contact for
more info

370. Solar Heat Corporation
1252 French Avenue
Lakwood, OH 44107
(216) 228-2993
Product Codes:
AHE   AHW   CFL   FEA
KIT

371. Solar Heater Manufacturer
1011 6 Ave. South
Lake Worth, FL 33460
(305) 586-3839
Product Codes:
AHW   APO   AHC   CFL
ASH   AGH   AMH   COS
STL   SYL   INS   HEX   KIT
NBS   LAB

372. Solar Heating of New Jersey
811 Wynetta Place
Paramus, NJ 07652
(201) 652-3819
Product Codes:
CFL   COC   AHW   AHE
APO   CCM   CMR

373. Solar Heating Systems
13584 49th Street North
Clearwater, FL 33520
(813) 577-3961
Product codes:
AHE   AHW   CFL   FEA
APO   COS   COC   KIT
CCM

374. Solar Home Systems Inc.
12931 West Geauga Trail
Chesterland, OH 44026
(216) 729-9350
Product Codes:

AHW   APO   AHE   CFL
CFA   COC   STA   SYA
SYL   NBS   ASH

375. Solar Hydro Inc.
765 S. State College Blvd.
Fullerton, CA 92631
(714) 992-4470
Product Codes:
AHW   AHE   APO   AAG
CFL   COS   STL   SYL   SPC
INS

376. Solar Inc.
P.O. Box 246
Mean, NE 68041
(402) 624-6555
Product Codes:
AHE   SYA   STP   COS
NBS   FEA

377. Solar Industries Inc.
Monmouth Airport Industrial Park
Farmingdale, NJ 07727
(201) 938-7000
Product Codes:
APO   SYL   CFL   COC

378. Solar Industries Ltd.
1727 Llewelyn
Baltimore, MD 21213
(301) 732-2072
Product Codes:
AHE   SSF

379. Solar Industries of Florida
P.O. Box 9013
Jacksonville, FL 32208
(904) 768-4323
Product Codes:
AHE   AHW   CFL   FEA

380. Solar Innovations
412 Longfellow Blvd.
Lakeland, FL 33801
(813) 688-8373
Product Codes:
AHW   SYL   AHE   APO
CFL   COS   INS   KIT   FEA
STL   CCM   CMM   CMP

381. Solar Kenetics Corp.
P.O. Box 17308
West Hartford, CT 06117
(203) 233-4461
Product Codes:
AHW   APO   AGH   AAG
CCN   CCM   COS   COC
STL   SYL   HEX   FEA
LAB   ASH

382. Solar Kinetics Inc.
147 Parkhouse St./P.O. Box 10764
Dallas, TX 75207
(214) 747-6519
Product Codes:
AHE   AHW   CCN   FEA

383. Solar King International
8577 Canoga Ave.
Canoga Park, CA 91304
(213) 998-6400
Product Codes:
AHE   AHW   CFL   SYL
COC   APO

384. Solar Life
404 Lippincott Ave.
Riverton, NJ 08077
(609) 829-7022
Product Codes:
AHW   SYL

385. Solar Living Inc.
P.O. Box 12
Netcong, NJ 07857

(201) 691-8483
Product Codes:
AHW   CFL   CCM   CMA
KIT

386. Solar Manufacturing Co.
40 Conneaut Lake Rd.
Greenville, PA 16125
(412) 588-2571
Product Codes:
AHE   CFA   CCM   STA
SYA   COS   CMG   SSF
FEA

387. Solar Northwest Corp.
Route 1/P.O. Box 114
Long Beach, WA 98631
(206) 642-2249
Product Codes:
AHW   COS

388. Solar One Ltd.
2644 Barret St.
Virginia Beach, VA 23451
(804) 340-7774
Product Codes:
AHE   STA   SYA   CFA
FEA

389. Solar Physics Corporation
1350 Hill Street, Suite A
El Cajon, CA 92020
(714) 440-1625
Product Codes:
AHC   CFL   CCN   FEA
AHW   APO

390. Solar Pool Heaters of SW Florida
901 S.E. 13th Place
Cape Coral, FL 33904
(813) 542-1500
Product Codes:
APO   CFL   FEA

391. Solar Power Inc.
201 Airport Blvd - Cross Keys
Doylestown, PA 18901
(215) 348-9066
Product Codes:
AHE   SSF

392. Solar Power West
709 Spruce St.
Aspen, CO 81611
(303) 925-4698
Product Codes:
APO   CFL   KIT

393. Solar Pre-Fab Ltd.
2625 S.E. Kelley St.
Portland, OR 97202
(503) 233-1652
Product Codes:
AHW   CFL   KIT

394. Solar Products Inc.
12 Hylestead St.
Providence, RI 90295
(401) 467-7350
Product Codes:
AHW   AHE   CFL   SYL

395. Solar Products Inc./Sun-Tank
614 N.W. 62nd Street
Miami, FL 33150
(305) 756-7609
Product Codes:
AHE   AHW   CFL   FEA
APO   KIT   COC   CCM
CMP

396. Solar Products Manufacturing Corp.
151 John Downey Dr.
New Britain, CT 06051
(203) 224-2164
Product Codes:
AHE   AHW   APO   CFL
CCM   CMG   CMP   COS

SYL   HEX   KIT   FEA
NBS

397. Solar Research
525 N. Fifth Street
Brighton, MI 48116
(313) 227-1151
Product Codes:
AHE   AHW   CFL   FEA
APO   CCM   COC   STC
HEX   KIT   CMG

398. Solar Research Systems
3001 Red Hill Ave.
Costa Mesa, CA 92626
(714) 545-4941
Product Codes:
APO   CFL   COS   SYL
FEA

399. Solar Room Company
Box 1377
Taos, NM 97511
(505) 758-9344
Product Codes:
AHE   AGH   KIT

400. Solar Sauna
Box 446
Hollis, NH 03049
Product Codes:
AHE   AGH   KIT

401. Solar Sensor System
4220 Berritt St.
Fairfax, VA 22030
Product Codes:
AHE   COC   AHW   APO
COS   INS

402. Solar Shelter Engineering Co.
P.O. Box 179
Kutztown, PA 19530

(215) 683-6769
Product Codes:
AHE   SSF   FEA

403. The Solar Store
No. 1 Solar Lane
Parker, SD 57053
(605) 648-3465
Product Codes:
AHE   SSF

404. Solar Sun Inc.
235 W. 12th St.
Cincinnati, OH 45210
(513) 241-4200
Product Codes:
AHE   AHW   APO   CFL
SYL

405. Solar Sunstill Inc.
15 Blueberry Ridge Rd.
Setauket, NY 11733
(516) 941-4078
Product Codes:
CCM   CMC   CMR

406. Solar Supply Inc.
9163 Chesapeake Dr.
San Diego, CA 92123
(714) 292-7811
Product Codes:
COS   COC

407. Solar Systems
26046 Eden Landing Road
Hayward, CA 94545
(415) 785-0711
Product Codes:
AHE   AHW   APO   CFA
CFL   CCM   COC   STL
STC   SYA   SYL   HEP
HEX   KIT   CMP   CMB

408. Solar Systems by Sundance
4815 S.W. 75th Ave.
Miami, FL 33155
Product Codes:
AHE   AHW   CFL   FEA

409. Solar Tec Corp.
8250 Vickers St.
San Diego, CA 92111
(714) 560-8434
Product Codes:
APO   AHC   AHW   CFL
FEA   CCN   SYL   HEP
NBS

410. Solar Technology Corp.
2160 Clay
Denver, CO 80211
Product Codes:
CFA   SYA   AHC   KIT
FEA   AGH   SSF

411. Solar Technology Inc.
3927 Oakclif Industrial Court
Atlanta, GA 30340
(404) 449-0900
Product Codes:
AHE   AHW   CFL   CCM
COC   COS   STL   SYL
CMP   FEA

412. Solar Therm
203 Point Royal Dr.
Rockwall, TX 75807
(214) 475-2201
Product Codes:
AHE   AHW   APO   CFL
CCM   CMM   KIT   CMA
CMG   STL   CMP   CMI
COC

413. Solar Unlimited
4310 Governors Drive
Huntsville, AL 35805

(205) 837-7340
Product Codes:
HEX

414. Solar Usage Now Inc.
450 E. Tiffin St./P.O. Box 306
Bascom, OH 44809
(419) 937-2226
Product Codes:
AHW   CFL   KIT

415. Solar Utilities
P.O. Box 1696/2850 Mesa Verde Dr.
Costa Mesa, CA 92626
(714) 557-7125
Product Codes:
APO   CFL

416. Solar Utilities of Nebraska - Sun
922 Lake St./P.O. Box 387
Gothenburg, NE 69138
(308) 537-7377
Product Codes:
AHE   CFA   KIT   COC
AHW   SSF   COS   CCM
CMB

417. Solar Water Heaters of New Port Richey
1214 US. Highway 19 North
New Port Richey, FL 33552
(813) 848-2343
Product Codes:
AHW   AHE   APO   AMH
CFA   CCM   CMG   CMA
CMH   CMS   CMM   CMT
CMC   CMR   COS   STL
SYL   KIT   FEA   NBS
LAB   ASH

418. Solar West
2711 Chicago Ave.
Riverside, CA 92507
(714) 684-1555

Product Codes:
APO   CFL

419. Solar West Inc.
3636 N Hazel #108/P.O. Box 892
Fresno, CA   93714
(209) 222-3455
Product Codes:
AHW   AHE   APO   CFL

420. Solar World Inc.
4449 N 12th St./Suite 7
Phoenix, AZ   85014
(602) 266-5686
Product Codes:
CFL   STL   COS   SYL
AHW   AHE   APO

421. Solar 1/Div. Stellar Industries, Inc.
7265 Commerce Drive
Mentor, OH   44060
(216) 951-6363
Product Codes:
AHW   AHE   APO   CFA
COS   COC   STA   SYA
INS   FEA   NBS   ASH

422. Solarator, Inc.
P.O. Box 277
Madison Heights, MI   48071
(313) 642-9377
Product Codes:
AAG   AHC   AGH   AMH
AHW   APO   CFL   SPC

423. Solaray Corp.
2414 Makiki Heights Dr.
Honolulu, HI   96822
(808) 533-6464
Product Codes:
AHW   APO   CFL   SYL
AHE   FEA

424. Solaray Inc.
324 S. Kidd Street
Whitewater, WI   53190
(414) 473-2525
Product Codes:
AHE   AHW   CFA   CFL
FEA   APO   STA   SYA
SYL   COS   CCM   CMA
CMH   CMP   CMB

425. Solarbeam Industries, Inc.
118 N. Almansor St.
Alhambra, CA   91801
(213) 282-8451
Product Codes:
AHE   AHW   CFL

426. Solarcell Corp.
1455 N.E. 57th St.
Fort Lauderdale, FL   33334
(305) 462-2215
Product Codes:
AHW   APO   CCN   COS
CCM   CMM   KIT

427. Solarcoa
21157 E. Spring Street
Long Beach, CA   90808
(213) 426-7655
Product Codes:
AHE   AHW   CFL   FEA
SYL   APO   COC   CCM
CMP   CMM

428. Solarequip
P.O. Box 21447
Phoenix, AZ   85036
(602) 267-1166
Product Codes:
AHE   AHW   APO   CFL
STL   SYL   CCM   CMM
CMP   KIT   COC

429. Solargenics, Inc.
9713 Lurline Ave.
Chatsworth, CA   91311
(213) 998-0806
Product Codes:
AHW   CFL   SYL   KIT
AHE   COS   CCM   CMM
CMP   HEX   STL   NBS
FEA

430. Solargizer Corportion
220 Mulberry Street
Stillwater, MN   55082
(612) 439-5734
Product Codes:
AHE   AHW   CFL   FEA
APO   SYL   KIT   CCM
CMP   STL

431. Solargy Systems/Vaughn Corp.
386 Elm St.
Salisbury, MA   81950
(617) 462-6683
Product Codes:
AHW   SYL   AHC   APO

432. Solarmaster
722-D West Betteravia Rd.
Santa Maria, CA   93454
(805) 922-0205
Product Codes:
CFL   CCN   KIT   AHC
AHW   APO   AGH   AAG

433. Solarmaster Systems, Inc.
20 Republic Rd.
N. Billerica, MA 01862
(617) 667-4668
Product Codes:
AHW   AHE   APO   CFA

434. Solarmetrics
23 Bridge St.
Manchester, NH   03101

(603) 668-3216
Product Codes:
COS   COC   STA   STL   INS
AHC   AHW   APO

435. Solaron Corporation
300 Galleria Tower-720 S. Colorado
Denver, CO   80222
(303) 759-0101
Product Codes:
AHE   CFA   FEA   AHW
CCM   COS   STA   SYA
HEX   CMP   NBS

436. Solarsystems Inc.
507 W. Elm Street
Tyler, TX   75702
(214) 592-5343
Product Codes:
AHC   AHW   APO   CFE
FEA   NBS

437. Solartech, Inc.
250 Pennsylvania Ave.
Salem, OH   44460
(216) 332-9100
Product Codes:
AHW   AGH   CFA   CCM
CMA   CMM   COC   STL
HEX   NBS   LAB   ASH

438. Solartech Systems Corp.
Box 591
Devine, TX   78016
(512) 663-4491
Product Codes:
AHE   AHW   APO   CFL
SYL   NBS

439. Solartherm
1640 Kalmia Rd. N.W.
Washington, DC   20112
Product Codes:
AHW   AHC   CCN

440. Solarway
P.O. Box 217
Redwood Valley, CA   95470
(707) 485-7616
Product Codes:
AHE   CCM   CMG

441. Solar-Eye Products, Inc.
1300 N.W. McNab/Bldgs. G&H
Fort Lauderdale, FL   33309
(305) 974-2500
Product Codes:
APO   AHW   AHE   CFL
SYL   COC   FEA

442. Solar-Thermics Enterprises, Ltd.
Highway 34 East/P.O. Box 248
Creston, IA   50801
(515) 782-8566
Product Codes:
AHE   SSF

443. Solastor/Energy Absorption Sys.
860 S. River Rd.
W. Sacramento, CA   95691
(916) 371-3900
Product Codes:
AHW   AHC   APO   CFL
STL   SYL   HEX   CCM
CMP   CMM   COS   COC

444. Solectro-Thermo, Inc.
1934 Lakeview Ave.
Dracut, MA   01826
(617) 957-0028
Product Codes:
AHW   AHE   APO   CCN
CCM   CMA   CMM   CMT
CMC   CMR   CMB   COS
STA   SYA   NBS   LAB

445. Solergy Inc.
70 Zoe St.
San Francisco, CA   94107

(415) 495-4303
Product Codes:
AHE   AHW   CFL   CCN
NBS   FEA

446. Soltex Corp.
1804 Afton St. - Lock Lane
Houston, TX   77055
(713) 782-4478
Product Codes:
AHE   AHW   CFL   FEA

447. Soltrax Inc.
720 Rankin Rd. N.E.
Albuquerque, NM   87107
(505) 344-3431
Product Codes:
AHW   APO   AHC   ACA
AAG   CFL   CCN   COS
COC   STL   SYL   KIT
LAB

448. Solus Inc.
P.O. Box 35227
Houston, TX   77035
(713) 772-6416
Product Codes:
AHW   AHC   APO   AAG
CFL   SYL   HEP

449. Sol-Aire/Energy Systems Inc.
2750 S. Shoshone
Englewood, CO   80110
(303) 761-4335
Product Codes:
APO   AHC   AHW   SYA
CFA   STA   NBS   LAB
ASH

450. Sol-Era Energy Systems
P.O. Box 651
Worthington, OH   43085
(614) 846-8594
Product Codes:

AHW   AHE   APO   CFL
STL   SYL   STA

451. Southeastern Solar Systems, Inc.
4705 Bakers Ferry Rd. S.W.
Atlanta, GA 30336
(404) 691-1864
Product Codes:
AHW   AHC   APO   CFL
FEA   CTU   CCN   CCM
STC   SYL   COS   STL   SYP
HEX   HEP   CMA   CMP
NBS   AAG

452. Southern Lighting/Universal 100
Products
501 Elwell Avenue
Orlando, FL 32803
(305) 894-8851
Product Codes:
AHE   AHW   CFL   FEA
COS   CCM

453. Southwest Air Conditioning Inc.
7268 El Cajon Blvd.
San Diego, CA 92115
(714) 462-0512
Product Codes:
AHW   AHE   APO   AMH
CFL   STL

454. Southwest Ener-Tech, Inc.
3030 S. Valley View Blvd.
Las Vegas, NV 89102
(702) 873-1975
Product Codes:
AHE   AHW   CFL   FEA

455. Southwest - Standard
P.O. Box 10094
El Paso, TX 79991
(915) 533-6291
Product Codes:
CFL   AHW   APO   NBS

456. Southwest - Standard
P.O. Box 14132
Albuquerque, NM 87111
(505) 265-8871
Product Codes:
CFL   AHW   APO   NBS

457. Spectran Instruments
P.O. Box 891
La Habra, CA 90631
(213) 694-3995
Product Codes:
AHC   AHW   INS

458. Spiral Tubing Corp.
533 John Downey Dr.
New Britain, CT 06051
(203) 244-2409
Product Codes:
AHE   AHW   CCM   CMA

459. SSP Associates
704 Blue Hill Rd.
River Vale, NJ 07675
(201) 391-4724
Product Codes:
AHE   AHW   CFL   CTU

460. Standard Electric Co.
P.O. Box 631
Rocky Mount, NC 27801
(919) 442-1155
Product Codes:
AHE   AHW   CFL   FEA

461. Standard Solar Collectors, Inc.
1465 Gates Ave.
Brooklyn, NY 11227
(212) 456-1882
Product Codes:
AHW   AHE   CFA   CFL
STL   SYA   SYL   KIT   FEA
ASH

462. State Industries, Inc.
Cumberland Street
Ashland City, TN 37015
(615) 792-4371
Product Codes:
AHE   AHW   CFL   FEA
SYL   STL   KIT

463. Stolle Corp.
1501 Michigan St.
Sidney, OH 45365
(513) 492-1111
Product Codes:
AHW   AHE   CFL   STL
SYL   HEX   NBS   ASH

464. Sun Century Systems
P.O. Box 2036
Florence, AL 35630
(205) 764-0795
Product Codes:
AHE   AHW   CFL   FEA
APO

465. Sun Chance
P.O. Box 506
South Fallsburg, NY 12779
(914) 434-6650
Product Codes:
AHW   CTU   CCM   CMT
CMP   COS

466. Sun God
P.O. Box 54
New Britain, PA 18901
(215) 368-7719
Product Codes:
AHW   CFL   COS   COC
HEX

467. Sun Harvesters, Inc.
211 N.E. 5th St.
Ocala, FL 32670
(904) 629-0687

Product Codes:
AHW   APO   AMH   AAG
STL   SYL   KIT

468. Sun Power Corp.
12785 S. E. Hiway 212
Portland, OR   97015
(503) 655-6282
Product Codes:
AHC   AHW   APO   CFE

469. Sun Power Northwest
16615 76th Ave./N.E.
Bothell, WA   98011
(206) 486-6632
Product Codes:
AHW   AHC   APO   CFA
CCM   CMC   CMB   COS
STA   SYA   HEX

470. Sun Power of Colorado
343 Van Gordan St./Bldg. 18/406
Lakewood, CO   80228
(303) 988-6200
Product Codes:
AHC   AHW   CFA

471. Sun Power Solar Engineering Co.
4032 Helix St.
Spring Valley, CA   92077
(714) 464-5322
Product Codes:
AHE   AHW   CCN   CCM
CMT

472. Sun Power Systems, Ltd.
1024 West Maude Ave./Suite 203
Sunnyvale, CA   94086
(408) 738-2442
Product Codes:
AHE   AHW   KIT   CFL
SYL   APO

473. Sun Saver Corp.
P.O. Box 276
North Liberty, IA   52317
(319) 626-2343
Product Codes:
AHE   APO   AAG   CFA
CCM   CMA   CMM   CMR
COS   COC   STA   STL
SYA   SSF   INS   HEX   KIT
FEA   NBS   LAB   ASH

474. Sun Stone
P.O. Box 941
Sheboygan, WI   53081
(414) 452-8194
Product Codes:
AHE   CFA   FEA   AHW
CCM   COS   COC   STA
STC   SYA   APO   STL
CMB   CMM

475. Sun Systems, Inc.
P.O. Box 347
Milton, MA   02186
(617) 268-8178
Product Codes:
AHE   AHW   CFL   FEA
APO

476. Sun Systems Inc.
P.O. Box 155
Eureka, IL   61530
(309) 467-3632
Product Codes:
AHW   CFA   COS   APO
STL   SYA   FEA   AHC
AAG   SSF

477. Sunburst Solar Energy Inc.
P.O. Box 2799
Menlo Park, CA   94025
(415) 327-8022
Product Codes:
AHE   AHW   APO   CFL

FEA   COS   STL   HEX
KIT   CCM   CMG   CMP
NBS   ASH

478. Suncraft Solar Systems
5001 East 59th Street
Kansas City, MO   64130
(816) 333-2100
Product Codes:
AHE   CFA   FEA   STA
SYA   COS

479. Sundevelopment Inc.
1108 Hanover Rd.
York, PA   17404
(717) 225-5066
Product Codes:
AHE   AHW   CFL

480. Sundog Solar
3800 N. Virginia St.
Reno, NV   89506
(702) 322-8080
Product Codes:
AHW   AHE   APO   AGH
AMH   CFL   CCM   CMA
CMM   COS   STL   SYL
INS   HEX   KIT   LAB

481. Sundu Company
3319 Keys Lane
Anaheim, CA   92804
(714) 828-2873
Product Codes:
APO   CFL   FEA   AHE
AHW

482. Sunearth Solar Products Corp.
RD 1 - Box 337
Green Lane, PA   18054
(215) 699-7892
Product Codes:
AHE   AHW   CFL   FEA
CCM   SYL   STL   CMA

289

CMG   CMH   CMS   CMM
CMP   COC   COS

483. Sunflower Energy Works Inc.
110 North Main
Hillsboro, KS   67063
(316) 947-5781
Product Codes:
CFA   AHE

484. Sunglaze
P.O. Box 2634
Olympic Valley, CA   95730
(702) 831-2400
Product Codes:
AHW   AHE   AAG   AGH
CFA   SYA

485. Sunhouse Inc.
6 Southgate Dr.
Nashua, NH   03060
(603) 888-0953
Product Codes:
AHC   AHW   CFL   SYL

486. Sunkeeper
Box 34 Shawsheen Village Station
Andover, MA   01801
(617) 470-0555
Product Codes:
AHW   COC   INS   AHC
COS

487. Sunpower Industries Inc.
10837 B6 S.E. 200th
Kent, WA   98031
(206) 854-0670
Product Codes:
AHE   CFA   SYA   KIT

488. Sunpower Systems Corporation
510 S. 52nd St./Suite 101
Tempe, AZ   85281
(602) 968-7425

Product Codes:
AHW   AHC   APO   CCN
FEA   SYL

489. Sunray Solar Heat Inc.
202 Classon Ave.
Brooklyn, NY   11205
(212) 857-0193
Product Codes:
AHW   AHE   APO   AMH
CFL   COS   SYL   KIT

490. Sunrise Solar Inc.
7359 Reseda Blvd.
Resdea, CA   91335
(213) 881-3164
Product Codes:
AHE   AHW   CFL

491. Sunsav Inc.
890 East Street
Tewksbury, MA   01876
(617) 851-5913
Product Codes:
AHC   AHW   CFL   FEA
CCM   CCN   STL   SYL
APO   ACO   ACA   CMG
CMA   CMM   CMC   CMP
CMB   COS   HEP   NBS
ASH

492. Sunsaver Inc.
P.O. Box 21672
Columbia, SC   29221
(803) 781-4962
Product Codes:
AHW   AGH   CFL   STL
SYL   HEX   KIT

493. Sunshine Greenhouses
109 Cooper St./Suite 5
Santa Cruz, CA   95060
(408) 425-1451

Product Codes:
AGH   AHW   APO   CFL

494. Sunshine Unlimited
900 North Jay St.
Chandler, AZ   85224
(602) 963-3878
Product Codes:
AHW   AHE   CFL

495. Sunshine Utility Company
1444 Pioneer Way, Suite 9 & 10
El Cajon, CA   92020
(714) 440-3151
Product Codes:
AHE   AHW   CFL   APO
FEA

496. Sunspot Environmental Energy Systems
P.O. Box 5110
San Diego, CA   92105
(714) 264-9100
Product Codes:
AHW   AHE   APO   CFL
SYL   KIT

497. Sunspot/Div. Elcam Inc.
5330 Debbie Lane
Santa Barbara, CA   93111
(805) 964-8676
Product Codes:
AHE   CFA   FEA   CCM
SYL   CFL   COC   AHW
STL   CMP

498. Suntap Inc./Bross Utilities Service, Inc.
42 East Dudley Town Rd.
Bloomfield, CT   06002
(203) 243-1781
Product Codes:
AHW   AHC   APO   AGH
AMH   AAG   CFL   CTU
CCM   CMH   CMI   COS
COC   STL   SYL   INS

290

HEX   FEA   NBS   LAB
ASH

499. Sunwall Inc.
P.O. Box 9723
Pittsburgh, PA   15229
(412) 364-5349
Product Codes:
AHE   AHW   APO   CFA
FEA   STA   STL   SYA
SYL   SYP   KIT

500. Sunwater Energy Products
1488 Pioneer Way/Suite 17
El Cajon, CA   92020
(714) 579-0771
Product Codes:
AHW   APO   CFL   COS
AGH   AHE   CCM   CMA

501. Sunworks Division/Enthone Inc.
Box 1004
New Haven, CT   06508
(203) 934-6301
Product Codes:
AHE   AHW   CFA   CFL
FEA   CCM   CMH   SYA
SYL   KIT

502. Sun-Dance Inc.
13939 NW 60th Ave.
Miami Lakes, FL   33014
(305) 557-2882
Product Codes:
AHE   AHW   APO   CCM
CMP   STL   SYL

503. Swan Solar
6909 Eton St./Unit G
Canoga Park, CA   91303
(213) 884-7874
Product Codes:
AHW   AHE   APO   CFL
HEX

504. Swedcast Inc.
7350 Empire Dr.
Florence, KY   41042
(606) 283-1501
Product Codes:
AHE   AHW   APO   CCM

505. Swedlow Inc.
12122 Western Ave.
Garden Grove, CA   92645
(714) 893-7531
Product Codes:
CCM   CMF

506. Systems Technology Inc.
P.O. Box 337
Shalimar, FL   32579
(904) 863-9213
Product Codes:
AHW   APO   CFL   CCM
SYL   CMP   STL   HEX
COS   COC   FEA

507. S.P.S. Inc.
8801 Biscayne Blvd.
Miami, FL   33138
(305) 754-7766
Product Codes:
AHC   CCN   CTU   CCM
CMT   RCE

508. Taco Inc.
1160 Cranston St.
Cranston, RI   02920
(401) 942-8000
Product Codes:
HEX   COS   COC   CCM
CMP

509. Technitrek Corp.
1999 Pike Ave.
San Leandro, CA   94577
(415) 352-0535
Product Codes:

AHW   APO   CFL   CCM
CMP   COS   COC   AHE
STL   SYL

510. Technology Applications Lab
1670 Highway A1A
Satellite Beach, FL   32937
(305) 777-1400
Product codes:
INS

511. Tektronix Products
P.O. Box 500
Beaverton, OR   97077
(503) 644-0161
Product Codes:
LAB

512. Temp-O-Matic Cooling Company
87 Luquer Street
Brooklyn, NY   11231
(212) 624-5600
Product Codes:
AHE   AHW   CFL   FEA

513. Texas Electronics Inc.
5529 Redfield St.
Dallas, TX   75209
(214) 631-2490
Product Codes:
INS

514. The Lord's Power Co. Inc.
726 Marshall St.
Albert Lea, MN   56007
(507) 377-1820
Product Codes:
AHE   CFA   SYA   KIT

515. Thermon Manufacturing Co.
100 Thermon Drive
San Marcos, TX   78666
(516) 392-5801
Product Codes:
CCM   CMA   AHE   AHW

516. Thomason Solar Homes Inc.
6802 Walker Mill Rd. S.E.
Washington, DC 20027
(301) 292-5122
Product Codes:
AHW   CFL   CCM   COC
STL   SYL   AHC

517. Thorton Sheet Metal
Waterboro, ME 04087
(207) 247-3121
Product Codes:
SSF

518. Tranter Inc.
735 East Hazel Street
Lansing, MI 48909
(517) 372-8410
Product Codes:
AHE   AHW   CCM   CMA
HEX

519. Tri-State Sol-Aire Inc.
7100 Broadway/Suite 6N
Denver, CO 80221
(303) 426-4000
Product Codes:
AHW   AHE   CFA   COS
STA   SYA   HEP

520. Troger Enterprises
2024 "A" De La Vina
Santa Barbara, CA 93105
(805) 687-6522
Product Codes:
AHE   AHW   APO   COS

521. Trol-A-Temp
725 Federal Ave.
Kenilworth, NJ 07033
(201) 245-3190
Product Codes:
AHE   AHW   COS

522. The Tub Company
Box 8
Boulder, CO 80306
(303) 449-4563
Product Codes:
AHE   AHW   SYL   KIT

523. Union Carbide Corp.
270 Park Ave.
New York, NY 10017
(212) 551-2261
Product Codes:
CCM   CMA

524. Unit Electric Control Inc./Sol-Ray Div.
130 Atlantic Drive
Maitland, FL 32751
(305) 831-1900
Product Codes:
AHE   AHW   CFL   FEA
COC   CCM   CMP   KIT

525. United Solar
Box 67
Steamboat Rock, IA 50672
(515) 868-2410
Product Codes:
AHE   CFA   SYA   COS
LAB   ASH

526. Unites States Solar Pillow
P.O. Box 987
Tucumcari, NM 88401
(505) 461-2608
Product Codes:
APO   CFL   FEA   SYP
SYA   SYL   CFA   AHC
AGH   AAG

527. Unitspan Architectural Systems Inc.
9419 Maon Ave.
Chatsworth, CA 91311
(213) 998-1131
Product Codes:

AHE   AHW   CFL   APO
KIT   FEA

528. Universal Solar Energy Company
1802 Madrid Avenue
Lake Worth, FL 33461
(305) 586-6020
Product Codes:
AHE   AHW   CFL   FEA
APO   CCM   COC   SYL
CMP

529. Universal 100 Products/Southern
Lighting
501 Elwell Ave.
Orlando, FL 32803
(305) 894-8851
Product Codes:
AHE   AHW   CFL

530. Urethane Molding Inc.
Route 11
Laconia, NH 03246
Product Codes:
AHE   AHW   CCM

531. Van Hussel Tube Corp.
Warren, OH 44481
(216) 372-8221
Product Codes:
CCM   CMA

532. Vanguard Solar Systems
2727 Coronado St.
Anaheim, CA 92806
(714) 871-8181
Product Codes:
AHW   APO   CFL

533. Vaughn Corp./Solargy
386 Elm St.
Salisbury, MA 01950
(617) 462-6683
Product Codes:
AHW   SYL   AHC   APO

534. Vertrex Corp.
108 Carlson Bldg. 808-106th N.E.
Bellevue, WA 98004
Product Codes:
AHE   COC

535. Vinyl-Fab Industries
930 E. Drayton
Ferndale, MI 48220
(313) 399-8745
Product Codes:
APO   SPC

536. Vinyl-Fab Industries
10800 St. Louis Dr.
El Monte, CA 91731
Product Codes:
SPC

537. Virginia Solar Components Inc.
Route 3/Highway 29 South
Rustburg, VA 24588
(804) 239-9523
Product Codes:
AHW   AHC   APO   CFL
CCM   CMP   STL   SYL
COS   COC   INS

538. W & W Solar Systems, Inc.
399 Mill St.
Rahway, NJ 07065
(201) 925-5488
Product Codes:
AHE   AHW   APO   CFL

539. Wallace Company
831 Dorsey Street
Gainsville, GA 30501
(404) 534-5971
Product Codes:
AHE   AHW   CFL   FEA
CCM   SYL   APO   HEP

540. Weathertronics
2777 Del Monte St.
West Sacramento, CA 95691
(916) 371-2660
Product Codes:
INS

541. Weather-Made Systems Inc.
West Hwy 266/Rt 7/Box 300-D
Springfield, MO 65802
(417) 865-0684
Product codes:
AHE   AHW   CFL

542. Weather-Made Systems Inc.
Rt 2/Box 268-S
Lamar, MO 64759
(417) 682-3489
Product Codes:
AHE   AHW   CFL

543. Weksler Instruments Corp.
80 Mill Road/P.O. Box 3040
Freeport, NY 11520
(516) 623-0100
Product Codes:
COC

544. Western Energy Inc.
454 Forest Ave.
Palo Alto, CA 94302
(415) 327-3371
Product Codes:
AHE   AHW   APO   CFL
COC   STL   HEX   CMP
CCM   NBS

545. Western Solar Development Inc.
1236 Callen St.
Vacaville, CA 95688
(707) 446-4411
Product Codes:
AHC   AHW   APO   CFL
KIT

546. Westinghouse Electric Corp.
5005 Interstate Drive North
Norman, OK 73069
(405) 364-4040
Product Codes:
HEP

547. Wilcon Corporation
3310 S.W. Seventh
Ocala, FL 32670
(904) 732-2550
Product Codes:
AHE   AHW   FEA   CFL
SYL   APO   COC   COS

548. Wilcox Manufacturing Corp.
P.O. Box 455
Pinellas Park, FL 33565
(813) 531-7741
Product Codes:
AHW   AHE   APO   CFL
FEA   HEX   HEP

549. Wilshire Foam Products Inc.
P.O. Box 34217
Dallas, TX 75234
(214) 241-4073
Product Codes:
SPC

550. Wilson Solar Kinetics Corp.
P.O. Box 17308
West Hartford, CT 06117
(203) 233-4461
Product Codes:
CCN   FEA   AHW   APO
COS   CMP   STL   SYL
AAG   CCM   CMM

551. Wojick Industries Inc.
527 N. Main St.
Fallbrok, CA 92028
(714) 728-5593

Product Codes:
APO   CFL   SYL   COC

552. Wolverine Tube Div./Universal Oil
Prod.
P.O. Box 2202
Decatur, AL   35601
(205) 353-1310
Product Codes:
HEX

553. Wormser Scientific Corporation
88 Foxwood Rd.
Stamford, CT   06903
(203) 322-1981
Product Codes:
AHC   CCN

554. N H Yates & Co. Inc.
227c Church Lane
Cockeysville, MD   21030

(301) 667-6300
Product Codes:
AHE   AHW   CFL   CCM
CMP

555. Ying Manufacturing Corp.
1957 West 144th Street
Gardena, CA   90249
(213) 327-8399
Product Codes:
AHW   CFL   FEA   AHC
APO   SYL   CFA   HEX
KIT   CCM   CMP   CMB
CMM   NBS   ASH   STL
STA

556. Zomeworks Industries
P.O. Box 712
Albuquerque, NM   87103
(505) 242-5354
Product Codes:

SYP   AHC   AHW   SYL
SYA   CCM   CMR   CMT
KIT   FEA

557. ZZ Corp.
10806 Kaylor St.
Los Alamitos, CA   90720
(213) 598-3220
Product Codes:
AHW   AHE   AGH   AMH
CCN   CCM   CMI   CMM
CMT   CMR   COC   STL
SYL   INS   FEA   NBS
LAB   ASH

558. 3-M Co.
Box 33331 Stop 62
St. Paul, MN   55133
Product Codes:
AHE   AHW   CCM   CMC
CMF

GEOGRAPHICAL LISTING

ALABAMA
124, 168, 201, 239, 413, 464, 552.

ARIZONA
39, 40, 41, 103, 106, 117, 170, 207, 218, 219, 343, 420, 428, 488, 494.

CALIFORNIA
5, 8, 9, 17, 18, 21, 27, 29, 31, 36, 38, 46, 49, 50, 52, 60, 65, 73, 74, 76, 87, 90, 96, 100, 113, 118, 119, 126, 133, 136, 150, 147, 163, 167, 171, 176, 179, 181, 182, 184, 185, 186, 195, 209, 211, 213, 222, 227, 247, 248, 254, 258, 259, 264, 267, 276, 278, 285, 286, 293, 297, 306, 316, 318, 319, 324, 327, 329, 332, 347, 348, 349, 350, 360, 361, 366, 375, 383, 389, 398, 406, 407, 409, 415, 418, 419, 425, 427, 429, 432, 440, 443, 445, 453, 457, 471, 472, 477, 481, 484, 490, 493, 495, 496, 497, 500, 503, 505, 509, 520, 527, 532, 536, 540, 544, 545, 55k, 555, 557.

COLORADO
23, 53, 104, 123, 129, 138, 141, 151, 155, 202, 236, 301, 334, 340, 353, 392, 410, 435, 449, 470, 519, 522.

CONNECTICUT
25, 54, 64, 128, 137, 150, 198, 200, 215, 242, 257, 265, 288, 289, 381, 396, 458, 498, 501, 550, 553.

DELAWARE
112, 183

FLORIDA
15, 28, 37, 45, 48, 56, 59, 67, 79, 91, 95, 98, 102, 142, 143, 154, 164, 172, 187, 223, 244, 260, 296, 299, 308, 314, 335, 337, 339, 342, 345, 346, 352, 355, 359, 368, 37k, 373, 379, 380, 390, 395, 408, 417, 426, 441, 452, 467, 502, 506, 507, 510, 524, 528, 529, 547, 548.

GEORGIA
197, 241, 243, 305, 357, 411, 451, 539

ILLINOIS
11, 12, 22, 77, 81, 105, 177, 206, 210, 229, 240, 246, 262, 271, 292, 302, 312, 317, 321, 338, 476.

INDIANA
42, 51, 295

IOWA
101, 125, 194, 204, 224, 255, 277, 442, 473, 525.

KANSAS
192, 225, 303, 367, 483

KENTUCKY
217, 235, 504

LOUISIANA
307

MAINE
111, 309, 341, 517

MARYLAND
2, 63, 220, 331, 358, 378, 554.

MASSACHUSETTS
88, 99, 107, 108, 115; 148, 165, 205, 216, 273, 320, 325, 326, 431, 433, 444, 475, 486, 491, 533.

MICHIGAN
6, 78, 110, 116, 130, 397, 422, 518, 535.

MINNESOTA
1, 190, 193, 238, 251, 351, 310, 363, 430, 514, 558.

MISSOURI
80, 233, 478, 541, 542.

NEBRASKA
323, 376, 416.

NEVADA
294, 454, 480.

NEW HAMPSHIRE
92, 169, 189, 212, 221, 249, 400, 434, 485, 530.

NEW JERSEY
10, 24, 30, 55, 61, 66, 83, 94, 97, 173, 174, 250, 304, 356, 365, 372, 377, 384, 385, 459, 521, 538.

NEW MEXICO
13, 313, 399, 447, 456, 526, 556.

NEW YORK
3, 7, 20, 34, 57, 75, 89, 114, 146, 162, 188, 191, 232, 281, 290, 405, 461, 465, 489, 512, 523, 543.

NORTH CAROLINA
69, 70, 71, 72, 120, 166, 330, 460.

OHIO
16, 139, 156, 158, 226, 234, 237, 256, 261, 268, 284, 328, 351, 369, 370, 374, 404, 414, 421, 437, 450, 463, 531.

OKLAHOMA
35, 58, 84, 231, 282, 546.

OREGON
214, 393, 468, 511.

PENNSYLVANIA
14, 19, 32, 33, 47, 68, 127, 131, 145, 153, 180, 199, 266, 270, 279, 280, 283, 298, 300, 315, 386, 391, 402, 466, 479, 482, 499.

RHODE ISLAND
132, 196, 394, 508.

SOUTH CAROLINA
157, 175, 354, 492.

SOUTH DAKOTA
403.

TENNESSEE
44, 121, 122, 208, 462.

TEXAS
4, 26, 62, 85, 109, 134, 149, 160, 161, 252, 269, 311, 333, 364, 382, 412, 436, 438, 446, 448, 455, 513, 515, 549.

UTAH
274.

VERMONT
152, 159, 322.

VIRGINIA
82, 93, 178, 203, 230, 263, 275, 291, 336, 344, 388, 401, 537.

WASHINGTON
135, 144, 228, 253, 387, 469, 487, 534.

WASHINGTON, D.C.
245, 439, 516.

WYOMING
272.

APPENDIX E

Bibliography

NON-TECHNICAL

F. Hickok, *The Buy Wise Guide to Solar Heat.* Hour House, St. Petersburg, Florida, 1976.

F. Daniels, *Direct Use of the Sun's Energy.* Ballantine Books, Westminster, Maryland, 1964.

I. Spetgang and M. Wells, *Your Home's Solar Potential.* Edmund Scientific, Barrington, New Jersey, 1976.

TECHNICAL

J. F. Kreider and F. Kreith, *Solar Heating and Cooling: Engineering, Practical Design, and Economics.* McGraw-Hill, New York, 1975.

ARCHITECTURAL

S. Braden III, *Graphic Standards of Solar Energy.* CBI, Boston, Massachusetts, 1977.

W.A. Schurcliff, *Solar Heated Buildings: A Brief Survey.* Cambridge, Massachusetts, 1977.

Directories

W. A. Schurcliff, *Informal Directory of the Organizations and People Involved in the Solar Heating of Buildings.* Cambridge, Massachusetts, 1976.

Solar Energy and Research Directory. Ann Arbor Science Publishers, Ann Arbor, Michigan, 1977.

C. W. Martz, *Solar Energy Source Book.* Solar Energy Institute of America, Washington, D.C., 1977.

GOVERNMENT PUBLICATIONS

F.E.A., *Buying Solar.* Stock No. 041-018-00120-4, Superintendent of Documents, Government Printing Office. Washington, D.C., 1972.

D. Barret et al, *Home Mortgage Lending and Solar Energy.* Stock No. 032-000-00387-2, Superintendent of Documents, Government Printing Office, Washington, D.C., 1977.

Index

Index